INTERMEDIATE
CALCULATOR
MATH

Gerardus Vervoort • Dale J. Mason

A FEARON
MAKEMASTER® BOOK

David S. Lake Publishers
Belmont, California

The Calculator Math Series:
Beginning Calculator Math (Book 1)
Intermediate Calculator Math (Book 2)
Advanced Calculator Math (Book 3)

Consultant: Alan Ovson
Cover designer: Susan True
Editorial and graphic services: Sheridan Publications Services,
Nevada City, California

Note: Beginning Calculator Math, Intermediate Calculator Math, and
Advanced Calculator Math are based on *Calculator Activities for the
Classroom 1, 2, and 3* (ditto master books), and *Calculator Activities
for the Classroom: Teacher's Resource Book,* published by Copp
Clark Publishing, a division of Copp Clark Limited, Toronto, Canada.
Some additional activity sheets may be found in the original
publications, which are available from Copp Clark Publishing,
517 Wellington Street West, Toronto, Ontario, Canada, M5V 1G1.

ISBN-0-8224-1201-2

Printed in the United States of America.

1.9 8 7 6 5 4

Preface

This comment by the inventor of logarithms still applies today. Manipulation of numbers is not the prime objective of an arithmetic program—in fact, it is only a minor part. People must know *when* to multiply and subtract as well as know *how* to multiply and subtract. They also must know what information is required to solve a problem and whether a proposed answer is reasonable.

Problem solving in the broadest sense is the essence of mathematics, and it is here that we find the calculator's value to the mathematics program. With the calculator eliminating the drudgery of lengthy calculations, the student can concentrate on the problem-solving process. Here are some specific situations where the calculator is useful:

— Estimating—mentally grasping the overall proportions of an operation—is an essential skill. The calculator enables students to make estimates and then rapidly get feedback on their correctness. Interestingly, while the calculator facilitates development of estimating skills, using a calculator also increases the need for that ability, for minor operating slips or use of a calculator with a low battery can easily produce incorrect answers. Users must thus develop a feeling for what the answer to a problem should be, and critically examine the answers that are displayed. *Elementary Calculator Math*, *Intermediate Calculator Math*, and *Advanced Calculator Math* each contains a series of activities to improve the students' ability to find approximate answers to addition, subtraction, multiplication, and division problems. Note that many educators differentiate between *estimation* and *approximation*, and the distinction may be useful. We have chosen to use the term *estimation* to cover both meanings. Those wishing to help their students learn both terms are encouraged to do so.

— Problem solving using mathematics is enhanced by use of the calculator. Students (and others) have often been dissuaded from attacking some types of problems because of the length of the computations involved. The calculator frees them to concentrate on what information is required to solve a problem, what steps and operations are involved, and what answers are reasonable. The *Intermediate* and *Advanced* books, in particular, show how both everyday problems and more intricate mathematical ones (too cumbersome for paper and pencil solution) can be handled easily with the aid of the calculator.

— Understanding of fractions can be enhanced by use of the calculator. It can also be used to develop the concept of a fraction and the relationship between common and decimal fractions. Calculator algorithms can be developed for operations with fractions. It should be noted that in spite of emphasis on the metric system and the increased emphasis on decimal fractions, common fractions have not disappeared entirely (nor should they). Not only is the concept of 1/3 much easier to grasp than $0.\bar{3}$, but students in the intermediate grades must be prepared to deal with algebra of the form $\frac{a + b}{c}$. Also, fractions provide opportunities for students to think their way through various calculator algorithms. The *Intermediate* and *Advanced* books contain many activities that use common and decimal fractions. A series of exercises is included to help students perform operations such as 3/4 + 5/8 on the calculator.

— Detecting and identifying patterns is the essence of inductive reasoning. The calculator can be the "number laboratory" where the student performs a series of numerical experiments, develops a hypothesis, and checks the conclusion by further experiments. In the past, these types of exercises, though recognized as desirable, were impossible due to the drudgery of the required computations. Many activities on patterns in all three *Calculator Math* books provide opportunities for students and teachers to engage in this type of experimentation.

— All teachers know that games, tricks, and puzzles help motivate students. They can be used to drill facts, expand concepts, and provide opportunities for discovery. Therefore, each book of *Calculator Math* contains a section of games and similar activities for individuals, small groups, or whole classes.

The *Calculator Math* books have been created to make connections between the calculator's capabilities, the learning of concepts in mathematics, and the application of those concepts and the calculator's capabilities to realistic problems. The flexible MAKEMASTER® duplicatable format enables any of the activity sheets to be used in multiple copies with minimal expense. The teacher's guide and skill-and-topic annotations on the activity sheets help connect the activities to the mathematics curriculum. Many of the activities suggested on the sheets can be continued with similar material provided by the teacher or invented by the students. We urge teachers to go further than we have, to build even more activities for their classes. They will enrich their mathematics program by doing so; they will help their students productively use one of the technological marvels of the twentieth century; and they will mightily please Mr. Napier's ghost besides!

The Authors

Note: Materials and literature on calculator applications in the mathematics program are rapidly proliferating. Readers who wish to pursue the subject in greater depth are advised to write the Calculator Information Center, 1200 Chambers Road, Columbus, Ohio 43212 (Telephone 614-422-8509). The center publishes a regular bulletin, free of charge, listing available materials and other resources.

Contents

Activities Overview

Activity Category:	Activity sheet numbers in:		
	Beginning Calculator Math	Intermediate Calculator Math	Advanced Calculator Math
Learning About the Calculator	1–10	1–8	1–5
Estimation	11–31	9–20	6–9
Games	32–42	21–28	10–15
Fractions		29–32	16–20
Patterns	43–44	33–34	21–30
Problem Solving	45–52	35–43	31–43
Content Lessons		44–47	44–46

Note: For ease of reference and to avoid the implications that *Beginning, Intermediate,* and *Advanced* may have for students, the teacher's notes and MAKEMASTER® activity sheets in the three *Calculator Math* books carry a double number. The first number indicates the book (*1-* denotes sheets in the *Beginning* book, *2-* sheets in the *Intermediate* book, and *3-* sheets in the *Advanced* book). The second number gives the activity-sheet number within that book. Thus, *2-14* indicates activity sheet 14 in the *Intermediate* book.

Topics-and-Skills Overview

Topics/Skills	Beginning Calculator Math (Book 1) Sheet numbers	Intermediate Calculator Math (Book 2) Sheet numbers	Advanced Calculator Math (Book 3) Sheet numbers
Calculator Mechanics			
familiarization with the calculator	1, 2, 3, 4, 5, 6, 7, 8, 10		1, 2, 3
speed and accuracy with the calculator	2, 3, 4, 5, 6, 7, 8, 10		
optimum use of the calculator		1, 2, 3, 4, 5, 6, 8	1, 2, 3, 4, 5, 18, 19 20, 35
judicious use of the calculator	9		32
special calculator algorithms	41		18, 19, 20, 44, 45, 46
Whole Numbers			
addition facts	2, 6, 7, 8, 10, 32, 37 38, 39	22	
subtraction facts	3, 6, 7, 8, 10, 33, 37 38, 39		
multiplication facts	4, 6, 7, 8, 10, 37, 38 39	1, 24, 26	10, 11
division facts	5, 6, 7, 8, 10, 37, 38 39, 40	24	10
order of operations	41	1, 8	
Estimation			
rounding	11, 12, 13, 14, 15, 16, 17−27, 28−31	10−21	6, 8, 9
estimation, addition	11, 12, 13, 14, 15, 16 48	9, 10, 11, 35	6
estimation, subtraction	17, 18, 19, 20, 34		6, 14
estimation, multiplication	21−27, 34, 35	12−19, 23, 25, 26, 36, 37, 41, 42	7, 8, 11, 13, 14
estimation, division	29, 30, 31, 35, 36	20, 21, 26	9, 11
large numbers		43	41, 42
using estimation to solve equations		45, 46	

Topics/Skills	Beginning Calculator Math (Book 1) Sheet numbers	Intermediate Calculator Math (Book 2) Sheet numbers	Advanced Calculator Math (Book 3) Sheet numbers
Common/Decimal Fractions			
concepts		29, 44	
equivalent fractions		29, 30, 31, 32	16, 17
converting, fractions/ decimals		29, 30, 31, 32	16, 17
operations with fractions	28		18, 19, 20
Problems			
using intuition		33, 34	
developing strategies	32, 33, 34, 35, 36, 38, 39		42
developing mathematical reasoning			30, 44, 45, 46
solving problems	45, 46, 47, 49, 51, 52	35, 36, 37, 38, 39, 40, 41, 42, 43	4, 5, 15
Patterns			
detecting/completing	33, 34, 35, 36, 43, 44	23, 33, 34, 44	21, 22, 23, 24, 25, 26, 27, 28, 29, 30
explaining		44	22, 23, 24, 25, 26, 27
Handling Data			
reading/constructing tables, graphs, diagrams	14, 47, 48	39, 40	31, 35, 36, 37
gathering/organizing data	45, 46	35, 37, 41, 42, 47	12, 15, 31, 34, 36, 37, 39
data and averages		22, 41, 42	12, 33, 34, 36, 37
Rate, Ratio, Percent			
rate, ratio, percent	49, 50, 51, 52	38, 41, 42	32, 34, 35, 36, 37, 38, 39, 40, 42
interest			1, 2, 4, 5
taxation			3, 31
consumer awareness	45, 46, 47, 49, 50	35, 36, 37, 38	1, 2, 3, 4, 5, 31, 32, 36

Topics/Skills	Beginning Calculator Math (Book 1) Sheet numbers	Intermediate Calculator Math (Book 2) Sheet numbers	Advanced Calculator Math (Book 3) Sheet numbers
Measurement			
linear measurement	49–52	38–42	
volume		38, 39, 41, 42	
area	49, 50, 51, 52	39, 40	
time		43	
Other Concepts			
number properties		5, 6, 7, 8, 44	18, 25, 26, 27
factors/multiples	36	24	10, 15
exponents, squares/square roots		4, 25	14, 22, 23, 24
limits			7, 28, 29, 39
pi		39, 47	40
negative integers		2	6
developing algebraic reasoning	42	27, 28, 45, 46	25, 26, 27, 43
inverse operations			44, 45, 46
significant digits	31	44	

Teacher's Guide

NOTE: Some Recommendations for Calculator Purchases

To many people, a calculator is a calculator. However, there are great differences between different brands and models of calculators. These differences include the type of logic employed, which dictates the order in which keys must be struck. Other differences are: available functions (%, $\sqrt{}$, and so on); size of display; automatic/manual constant; floating/fixed decimal point; automatic display shut-off; size; spacing and arrangement of keys; sturdiness of case; durability; and power source.

Selection of a particular model depends on the grade levels of the students who will be using the machine, and on the purposes of the program. Features that seem mandatory for calculators intended for intermediate school students include: sturdiness; algebraic mode (natural order arithmetic); floating decimal point; 8-digit display; automatic constant.

Rechargeable batteries are (or were) a frequent cause of breakdown. Moreover, depending on the recharging mechanism, the constant connection of rundown units to an external power source can be a nuisance. On the other hand, replacement batteries can be very costly in the long run.

Due to the large number of variations among calculators, it is most desirable that all students in the elementary classroom use the same model.

Eight Activity Sheets for Learning About the Calculator*

2-1 The Clever Calculator: Part I

Focus on . . .

- *Introducing the constant feature*
- *Using the calculator optimally*
- *Reviewing multiplication facts*

Notes

1. Check the constant feature of the calculator your students are using. Some models, such as the Texas Instrument 1200 (TI-1200), will display $\boxed{2}$ after pressing $(2)(+)(=)$; other models show $\boxed{4}$. The same TI-1200 will display $\boxed{4}$ after pressing $(2)(\times)(=)$. This inconsistency may confuse the students. Also, in some calculators the second factor is taken as the constant. These models will display $\boxed{18}$ and not $\boxed{12}$ after pressing $(2)(\times)(3)(=)(=)$.

2. Have the students start at zero and count by 2s to 100 using the calculator for immediate reinforcement. Instruct the students to push $\boxed{2}$ and $\boxed{+}$. Then have them say quietly to themselves the next number (4) and then push the $\boxed{=}$ key. The correct answer is displayed. The student goes on to say the next number (6) and pushes the $\boxed{=}$ key for immediate feedback, and so on. This exercise should be repeated by counting by 3s and 4s. Once this exercise has been done orally, the students should repeat it, filling in the blanks on the sheet before rechecking with the calculator.

Extension

Have students skip-count, starting from a number other than zero. For example, to count by 5s to 103, the student pushes $\boxed{3}\boxed{+}\boxed{5}$. The student then says the answer (8), then pushes the $\boxed{=}$ key for immediate reinforcement. The student then says the next number (13) and pushes the $\boxed{=}$ key again, and so on. This exercise can be repeated for any number.

*For solutions, see *Answer Key*, pp. 60–63.

2-2 The Clever Calculator: Part II

Focus on . . .

- *Using the constant function*

Notes

1. Students will need to know the meaning of *multiple*.
2. Subtraction with constants naturally leads into negative numbers. If the students have not studied these before, you will want to explain negative numbers as points on the number line, "to the left of the zero," without, at this stage, going into operations with negative numbers.
3. Similarly, division with constants leads to decimals.

Extension

After the students have completed the activity sheet and have had time to tinker with the constant function, discuss where and when the constants may be used; for example, making multiplication tables, skip counting, and so on.

2-3 The Clever Calculator: Part III

Focus on . . .

- *Using the constant function*

Note

Review the meaning and rules for the decimal point. For Exercise 6, students will need to round their answers in parts *b*, *d*, and *e*. Depending on their background, you can have them round to the nearest whole number, tenth, hundredth, or thousandth.

Extension

Discuss some consumer and business uses for the constant function such as sales tax, mark-up, discount, and so on.

2-4 Who's a Cube?

Focus on . . .

- *Learning about exponents*
- *Optimizing the use of the calculator*

Note

Discuss exponential notation, pointing out that the exponent tells how many times the number is used as a factor. Thus, 9^5 (read "nine to the fifth power") means 9 is used as a factor 5 times, $9 \times 9 \times 9 \times 9 \times 9$.

Extension

Have students find the powers of 10 from $10^1 = 10$ to $10^7 = 10,000,000$. Then show them how to write large numbers in *scientific notation*, starting with numbers such as $4,000,000 = 4 \times 10^6$ and then including numbers such as $137,000 = 1.37 \times 10^5$. This discussion can be expanded, depending on the ability and interest of the students.

2-5 What Multiplication Is: Part I
2-6 What Multiplication Is: Part II

Focus on . . .

- *Understanding multiplication*
- *Optimizing the use of the calculator*

Note

Most students should be familiar with this type of material. Emphasize the relation between addition and multiplication.

2-7 The Commutative Property

Focus on . . .

- *Reinforcing the concept of the commutative property*

Note

Have the students set up arrays to illustrate pairs of multiplication facts. For example,

```
4 × 7                      7 × 4

X  X  X  X        X  X  X  X  X  X  X
X  X  X  X        X  X  X  X  X  X  X
X  X  X  X        X  X  X  X  X  X  X
X  X  X  X        X  X  X  X  X  X  X
X  X  X  X
X  X  X  X
X  X  X  X
```

Write each of these as a number sentence.

$7 + 7 + 7 + 7 = $ _____

$4 + 4 + 4 + 4 + 4 + 4 + 4 = $ _____

Repeat the idea, using different arrays. Students should realize that the *order* in which two numbers are multiplied does not change the product.

2-8 The Important Order

Focus on . . .

- *Discovering the need for operation-order rules*
- *Optimizing the use of the calculator*

Notes

1. Have the students do part *a* on the activity sheet, checking to see whether their answers agree with the answers shown. Then have them work part *b*.
2. Discuss the difference between the problems in parts *a* and *b*. Students should notice that the problems in part *b* are rewritten so that the multiplication is done before the addition. Have students complete the sheet, emphasizing that they should multiply before they add.
3. Explain that when there are no parentheses to show which operation should be done first, the order for the operations is multiplication, division, addition, subtraction.

*Twelve Activity Sheets on Estimation**

2-9 Careful Comparisons

Focus on . . .

- *Estimating in addition*
- *Seeing number relations*

Notes

1. If necessary, review the meaning of > and <.
2. The long-term goal of this exercise is to improve the students' ability to estimate as a result of knowing whether the exact answer is somewhat greater or less than the answer obtained by rounding. It may be necessary to work through one or two more problems, pointing out the effects of rounding. For example, in the second problem, show that 244 is *4 more* than 240, and 127 is *3 less* than 130. Thus, 244 + 127 is *1 more* than 240 + 130.

2-10 Triple Choice

Focus on . . .

- *Estimating three-digit numbers*
- *Rounding to the nearest ten or hundred*

Note

The sums in each problem are sufficiently different for students to find the order by rounding the addends to the nearest hundred. However, if students have not had much practice in rounding, it may be worthwhile to have them round the addends to the nearest ten.

2-11 The Educated Guess

Focus on . . .

- *Estimating in addition*

Notes

1. The students may want to use their calculators immediately; have them write down their estimates first.
2. When the students have completed the page ask how many estimated too high the answers to Problem 1. Ask these students to describe how they arrived at their estimate. Then ask students who estimated too low to describe the method they used. Discuss the methods described, pointing out that for estimation to be useful, it must be both quick and reasonably accurate. Repeat the discussion for Problem 2, helping students to find an estimation method that works best.

2-12 Tri-Option

Focus on . . .

- *Estimating in multiplication*
- *Rounding to the nearest ten*

Notes

1. Have students write the rounded factors and products before selecting the largest product.
2. After students have checked their answers with their calculators, discuss any problems where their estimates were wrong. Ask them for suggestions on how they could improve their estimates.

2-13 Home, Home on the Range: Part I
2-14 Home, Home on the Range: Part II
2-15 Home, Home on the Range: Part III
2-16 Home, Home on the Range: Part IV
2-17 Home, Home on the Range: Part V
2-18 Home, Home on the Range: Part VI

Focus on . . .

- *Estimating in multiplication*
- *Developing and using the concept of a range*
- *Judging error in the estimation process*

Notes

1. Students must be competent at rounding two- and three-digit numbers to the nearest ten and hundred.

*For solutions, see *Answer Key*, pp. 64–69.

2. Note that the subtraction asked for in columns *d* and *e* to find the score requires the *absolute value* of the exact answer minus the estimate. The *lowest* score is the best.
3. Encourage students to narrow the range as much as possible. Discuss various ways in which this can be done. After each sheet in this series is completed, ask students whether they discovered any new way to narrow the range.

2-19 Somewhere In-between

Focus on . . .
- *Estimating in multiplication*
- *Estimating in division*

Notes
1. Students may have some initial difficulty with this sheet. Work through several problems as examples before having students complete the sheet.
2. The answers given, as with all estimation sheets, represent some of the possible answers a student may give.

Extension

Have students write different multiplication problems for each of the given ranges.

2-20 A Santa Claus Riddle

Focus on . . .
- *Estimating in division*

Extension

Have each student write three division problems whose quotients are almost the same, such as 371 ÷ 22, 86 ÷ 5, and 230 ÷ 14. Then ask them to find an estimation method that would give them the largest quotient.

Eight Activity Sheets with Games

2-21 The Guess and Check Game

Focus on . . .
- *Estimating in division*
- *Refining estimation abilities for accuracy*

Note

Discuss rules for rounding and decide whether actual answers will be rounded into integers, or to one or two decimal places. Stress that the students should estimate their answers, not calculate them with pencils and paper.

2-22 The Covered Target Game

Focus on . . .
- *Reinforcing addition facts*
- *Averaging*

Notes
1. One calculator for every two students is optimal.
2. The calculator displays can be covered with the flaps provided, or with wide rubber bands.
3. This game works well as a whole-class activity, as an activity station, or as a spare time game. The calculator allows the game to be virtually self-checking, although you may want to spot check the scoring.
4. The initial scores of 10 points each are given in order to make it less likely that one student will win by being silent. This way, each low score will improve the average.

Extensions

Once the students master the game, let them try these variations.
1. Make the target larger, and add two-digit numbers.
2. Make the target larger, and multiply by single-digit numbers.
3. Make the target zero, start from 100, and subtract single-digit numbers.
4. Make the target one, start from 1000, and divide by single-digit numbers.

2-23 Monstrous Multiplication

Focus on . . .
- *Estimating in multiplication*
- *Detecting patterns*

Notes
1. Provide envelopes for card storage. Have scissors available for cutting. For a more durable game, glue a piece of cardboard to the back of the numbers before cutting.

2. After sufficient playing time, have the students discuss strategy for a given set, such as 3, 4, 5, 6, 7, 8. They will see that the 8 and the 7 should go into the 100's place, the 6 and 5 into the 10's place, and the 4 and 3 into the unit's place.

Compare:
$$864 \times 753 \qquad 854 \times 763 \qquad 853 \times 764$$

Note that the sum is the same in all cases. Have the students see that if the sum of two digits, for example, is 11, then: $6 \times 5 > 7 \times 4 > 8 \times 3 > 9 \times 2$. *The rule is:* if the sum is constant, the closer numbers are together, the greater the product. It then follows that:

$$853 \times 764 > 854 \times 763 > 863 \times 754 > 864 \times 753$$

Generalize the pattern, given any set consisting of an even number of digits.

Extension

After the students have become very familiar with the original game, extend it to a set with an odd number of digits.

2-24 Divide and Conquer

Focus on . . .

- *Reviewing multiplication and division*
- *Factoring and multiples*

Notes

1. Review the meaning of factor and multiple. Be sure the students realize that: multiple ÷ factor = factor.
2. If a more durable spinner is needed, glue the faces onto a piece of plywood, drive a common nail into the center, and use a hairpin for a pointer.

2-25 Roundabout Route to a Square Root *

Focus on . . .

- *Estimating in multiplication or division*
- *Using square roots*

Notes

1. Review the meaning of square root.
2. To make the cards more durable, glue them onto heavier paper before cutting them.

Extension

Students may make additional cards by randomly punching two- or three-digit numerals followed by $\boxed{\times}$ $\boxed{=}$ to generate perfect squares.

*For solutions, see *Answer Key*, p. 70.

2-26 Destination 500

Focus on . . .

- *Reviewing multiplication facts*
- *Estimating in multiplication and division*

Note

Have students take turns choosing the destination number.

2-27 Two Astounding Tricks

Focus on . . .

- *Developing algebraic reasoning*

Extension

Ask students to explain why the tricks always work. For the first trick, they should use a, b, and c for the three numbers. After they perform all the operations, they should end up with $100a + 10b + c$. This will be a three-digit number with the digits in the original order. For the second trick, students should use x and $x + 1$ for the two numbers. When they perform the operations, the answer will be 5.

2-28 Faster Than a Speeding Calculator

Focus on . . .

- *Developing algebraic reasoning*
- *Discovering number relationships*

Extension

Students can make up problems similar to these using a number such as 900 instead of 999.

Four Activity Sheets on Fractions *

2-29 A Fraction by Another Name

Focus on . . .

- *Understanding the concept of fractions*
- *Converting fractions to decimal form*

Notes

1. Remind students that $1/4 = 1 \div 4$; $4/8 = 4 \div 8$, and so on.
2. Be sure students are familiar with the notation of using a bar over the digits that repeat in a repeating decimal ($1/3 = 0.\overline{3}$).

*For solutions, see *Answer Key*, pp. 71–73.

Extension

This sheet can be used as an introduction to equivalent fractions. Exercises *c* and *d* show that 1/3 = 2/6 because both equal $0.\overline{3}$. Students can be challenged to find other fractions equivalent to 1/3 and can check their answers on their calculators. They can then be asked to write a rule for finding equivalent fractions.

2-30 The Fraction Line

Focus on . . .

- *Converting fractions to decimal form*
- *Comparing fractions*

Notes

1. Students should change the fractions to decimals before placing the points on the number line.
2. This sheet prepares students for comparing fractions. Discuss the idea that one fraction is greater than a second fraction if it is to the right of it on the number line. Thus, 3/5 is greater than 4/7. Encourage students to try to find ways of deciding when one fraction is greater than another. This idea is considered in more detail on sheet 2-31.

Extension

Students may be encouraged to find other common fractions, labeling them E, F, G, H, . . . , to fit on the number lines.

2-31 Larger Fractions, Smaller Fractions

Focus on . . .

- *Comparing fractions*
- *Converting fractions to decimal form*

Note

There are many different ways of comparing fractions. Discuss these methods with students before they begin this sheet.

1. Inspection: $5/9 < 5/7$

   The numerators are the same, but ninths are smaller than sevenths.

2. Estimation: $3/5 > 1/2$
   $1/2 > 4/9$
   so: $3/5 > 4/9$

3. Converting to fractions with common denominators:

   $3/4 = 15/20$
   $4/5 = 16/20$
   $16/20 > 15/20$
   so: $4/5 > 3/4$

4. Algorithm using cross products:

   $$27 \qquad 20$$
   $$9 \times 3 \quad 5 \times 4$$
   $$\frac{3}{5} \diagtimes \frac{4}{9}$$
   $$27 > 20$$
   so: $3/5 > 4/9$

5. Converting to decimals:

   $3/4 = 0.75$
   $4/5 = 0.80$
   $0.80 > 0.75$
   so: $4/5 > 3/4$

Able students should be encouraged to use the method most suited to the particular question, rather than relying on only one method.

2-32 A Fraction by Another Name: II

Focus on . . .

- *Converting fractions to decimal form*
- *Building a table*
- *Comparing processes for determining decimal equivalents*

Notes

1. Students should have completed activity sheets 2-29 through 2-32 before attempting this sheet.
2. There are two methods for finding the decimal form of a fraction. 2/6, for example can be computed by using the calculator with multiplication:

$$\boxed{1}\; \boxed{\div}\; \boxed{6}\; \boxed{\times}\; \boxed{2}\; \boxed{=}\; \boxed{0.3333332}$$

Or, 2/6 may be computed by dividing:

$$\boxed{2}\; \boxed{\div}\; \boxed{6}\; \boxed{=}\; \boxed{0.3333333}$$

The difference in the answers is a result of the calculator's only going to eight digits. The effect is called "truncation." For most purposes the deviation is trivial.

Two Activity Sheets on Patterns*

2-33 Guess How It Goes From Here

Focus on . . .

- *Detecting and completing patterns*
- *Using intuition*

Notes

1. Review changing fractions to decimals.
2. Have students estimate the answers to the fourth, fifth, and sixth questions in each problem before trying them with their calculators.

Extension

Students can make up patterns of their own, using the ideas shown here.

2-34 Pattern Discovery

Focus on . . .

- *Detecting and completing patterns*
- *Drawing conclusions*

Notes

1. To complete Problem 2, students must understand operations with negative numbers.
2. Discuss the results of Problems *1b*, *1c*, and *3*. Point out that in the array with 10 numbers in a row, the products differ by 10; in the array with 6 numbers in a row, the products differ by 6. In the array with 10 numbers, the subtraction answers are 9 and 11, the counting numbers before and after 10. The result with 6 numbers is similar.

Extension

Challenge your better students to change the array to rows of eight numbers, rows of seven numbers, and so on, and see if they can discover whether the patterns hold. Other arrays may be made using only even numbers, negative numbers, or fractions. Have students think of more ways and experiment with various arrays.

Nine Activity Sheets on Problem Solving*

2-35 A Big Fish Story

Focus on . . .

- *Estimating in addition*
- *Gathering and organizing data*
- *Solving problems*

Note

You can have your students make "fishing trip lists" and shop from hardware catalogs that they supply. Students can also bring in other types of catalogs and shop for camping, hiking, or bicycling trips.

Extension

Discuss sales tax and shipping charges for different catalogs. Have students include these charges in the lists of items purchased.

2-36 Stocking Up

Focus on . . .

- *Estimating in multiplication*
- *Building consumer math skills*

Note

Discuss the various estimates made by the students, finding out who had the closest estimates and what estimating methods were used.

Extension

Students may be interested in planning the shopping lists for their imaginary trips. Advertisements from the local newspaper or a trip to the neighborhood supermarket will give the prices needed.

2-37 An Expensive House!

Focus on . . .

- *Estimating in multiplication*
- *Gathering information*
- *Building consumer math skills*

Extension

Have students work in groups of three or four to make lists of the materials needed to build a dog house. They should decide on the specifications and make sketches showing the dimensions of the finished building. Then have them "shop" for the materials at a local lumber yard or use catalogs.

2-38 Finding the Best Buy

Focus on . . .

- *Working with volume and ratio*
- *Calculating cost per unit*

Note

If necessary, review metric units and the formula for volume. Before the students start to work, have the whole class estimate the best buy.

*For solutions, see *Answer Key*, pp. 73—77.

2-39 The World's Largest Swimming Pool

Focus on . . .

- *Working with area and volume*
- *Gathering information from drawings*
- *Working with pi*

Notes

1. Before attempting these problems, students must be familiar with the formulas for the areas of rectangles, triangles, trapezoids, and circles, and the volumes of prisms and cylinders.

   Rectangle: $A = \ell \times w$
   Triangle: $A = \frac{1}{2} bh$
   Trapezoid: $A = \frac{1}{2} h (b_1 + b_2)$
   Circle: $A = \pi r^2$

   Box: $V = \ell \times w \times h$
   Cylinder: $V = \pi r^2 h$
   Prism: $V = Bh$ where B = Area of base

2. In each case, the volume is the area of one face (the bottom of the pool, the end wall of the tent) times the distance to the congruent face.
3. The tent problem is difficult and students may require extra help.
4. 355/113 is a better approximation of π than 22/7.

2-40 Brick Trick

Focus on . . .

- *Gathering data*
- *Constructing diagrams*
- *Working with area*

Notes

1. Make sure students know the formula $A = \ell \times w$.
2. Assign a student to find the cost of one brick.
3. Students may work in pairs.
4. Masking tape can be used to mark off one square meter, though this can slow down the exercise.
5. Students should be encouraged to draw and label diagrams of the wall under study.
6. Extra long tape helps in measuring a long wall.
7. Discuss the variations in students' answers.

Extensions

1. Students often like to look at other sets of building plans before drawing their own.
2. More capable students can design walls containing windows and doors.

2-41 A Drop in the Bucket: Part I
2-42 A Drop in the Bucket: Part II

Focus on . . .

- *Gathering data*
- *Working with volume calculations*
- *Using sampling and averages*
- *Estimating in multiplication*

Notes

1. This activity is designed for a few independent students, working in pairs.
2. Students must be capable of computing the volume of a figure with a cross section in the shape of a trapezoid.
3. Assure that students will have use of a sink for at least half an hour.

Extension

Pairs of students may want to compare their results, and discuss the reasons for inconsistencies.

2-43 Superstar Publicity

Focus on . . .

- *Working with large numbers*

Extension

Have students find other large numbers and make up problems similar to the ones on this sheet. For example, the length of time and number of steps needed to walk around the earth at the equator.

Four Activity Sheets with Content Lessons*

2-44 The Point in the Right Place

Focus on . . .

- *Working with rules for decimal placement*
- *Detecting and explaining patterns*

Notes

1. Have the students complete Exercises 1 and 2 with their calculators.
2. Compare the results of Exercise 1 with the results of Exercise 2.
3. Write the results of Exercise 2 on the board and discuss the answers. Examine each problem and see if the students can predict where the decimal point will go.

*For solutions, see *Answer Key*, pp. 78–79.

2-45 Solving Sentences: Part I
2-46 Solving Sentences: Part II

Focus on . . .

- *Solving equations with one variable*
- *Applying estimating skills*

Notes

1. Review solving equations in one variable by trial and error when the answer is a whole number.
2. Discuss solving an equation using trial and error when the answer is a decimal. Then work through the example on the activity sheet.

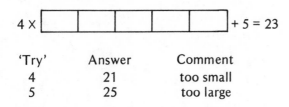

$$4 \times \boxed{} + 5 = 23$$

'Try'	Answer	Comment
4	21	too small
5	25	too large

Think—the answer must be between 4 and 5.

4.4	22.6	a bit too small
4.5	23	perfect

Extension

More capable students may find it interesting to make up code messages using problems similar to those on sheet 2-46.

2-47 The Pi Discovery

Focus on . . .

- *Discovering the relationship between circumference and diameter*
- *Gathering data*
- *Putting similar results in tabular form*

Notes

1. Collect a variety of circular objects: wheels, discs, cans, bottles, or plates. Provide rules, cloth tapes, and pocket calculators.
2. Divide the class in pairs.
3. Have the students measure the circumference and diameter of a variety of objects, and record their findings in the table.
4. Using their calculators, the students should complete the fourth column on the activity sheet: circumference ÷ diameter.
5. Discuss the similarity of the answers in column 4. Have the students compute their averages for column 4. Then, compute the class average for column 4.
6. Discuss the meaning of π, and ask students to try to define it.

Extension

Use π in a variety of ways. For example, if a bicycle tire has a diameter of 70 cm, how far will it roll in one revolution? In 10 revolutions? In 100 revolutions? How many revolutions would be required to go around the world?

2-1

The Clever Calculator: Part I

Skip-count fast—or faster

1. Press: ② ⊞ ⊟ ⊟ ⊟ ⊟ ⊟

Write your answers: ___ ___ ___ ___ ___

What's happening? _____

2. Press: ③ ⊞ ⊟ ⊟ ⊟ ⊟ ⊟ ⊟

Write your answers: ___ ___ ___ ___ ___ ___

What's happening? _____

3. Press ④ ⊞

What answers will be displayed if you press ⊟ ⊟ ⊟ ⊟ ⊟ ?

Answers: ___ ___ ___ ___ ___   Check!

4. Count by 5's, using your calculator. Fill in the blanks.

5, 10, ___, ___, ___, ___, ___, ___, ___, ___, ___, ___, ___, ___, ___, ___

5. Fill in the blanks. Check with your calculator.

6, 12, 18, ___, ___, ___, ___, ___, ___, ___, ___, ___, ___, ___, ___

9, 18, ___, ___, ___, ___, ___, ___, ___, ___, ___, ___, ___, ___

7, 14, ___, ___, ___, ___, ___, ___, ___, ___, ___, ___, ___, ___

20, 40, ___, ___, ___, ___, ___, ___, ___, ___, ___, ___, ___

8, 16, ___, ___, ___, ___, ___, ___, ___, ___, ___, ___, ___

50, 100, ___, ___, ___, ___, ___, ___, ___, ___, ___, ___, ___

• *Introducing the constant feature*
• *Using the calculator optimally*
• *Reviewing multiplication facts*

Intermediate Calculator Math

2-2

The Clever Calculator: Part II

How the constant helps you

1. **a.** Clear the calculator and press [2] [+] .
Now press [=] three times. (Don't clear.)
Write your answers. _____ , _____ , _____ .
What is happening? _____
_____ .

c. Clear the calculator and press [2] [×] [3] .
Now press [=] three times. (Don't clear.)
Write your answers. _____ , _____ , _____ .
What is happening? _____
_____ .

b. Clear the calculator and press [8] .
Now press [−] [2] .
Press [=] three times. (Don't clear.)
Write your answers. _____ , _____ , _____ .
What is happening? _____
_____ .

d. Clear the calculator and press [3] [2] [÷] [2] .
Now press [=] three times. (Don't clear.)
Write your answers. _____ , _____ , _____ .
What is happening? _____
_____ .

Your calculator has an "automatic constant" feature. In each of the preceding examples "2" was the "constant" which was repeatedly added, subtracted, multiplied, or divided.

2. Count by 4's to 100 by pressing [4] [+] , then [=] as many times as necessary. Fill in the blanks.

4, 8, 12, _____ , 20, 24, _____ , 32, 36, 40, 44, _____ , _____ , 56, 60, 64, _____ , _____ , _____ , 80,

_____ , 88, 92, _____ , 100.

3. Display the first ten multiples of 6.

6, _____ , _____ , _____ , _____ , _____ , _____ , _____ , _____ , _____ .

• *Using the constant function*

Name_____

Intermediate Calculator Math

2-3

The Clever Calculator: Part III

More about constants and the calculator

1. Push [6] [+] [=] . (Don't clear.)

Now push [4] [=] .

Answer:_____

(Don't clear.) Push [7] [=] .

Answer: _____

(Don't clear.) Push [8] [=] .

Answer: _____

What's happening? _____

_____ .

2. Add 3 to each of the following:
(Push [3] [+] [=] first. Don't clear.)

a. 7 _____

b. 24 _____

c. 5 _____

d. 9 _____

e. 42 _____

3. Add 623 to each of:

a. 94 _____

b. 165 _____

c. 947 _____

d. 382 _____

e. 6493 _____

4. Use a similar method to subtract 84 from each of the following:
(Push [1] [5] [6] [−] [8] [4] [=] first, then [9] [3] [7] [=]...)

a. 156 _____

b. 937 _____

c. 2518 _____

d. 36,945 _____

e. 18,763 _____

5. Multiply 1.07 by each of:

a. 36 _____

b. 945 _____

c. 167 _____

d. 25.84 _____

e. 3765.93 _____

6. Divide each of the following by 63:

a. 441 _____

b. 227 _____

c. 63 _____

d. 97.4 _____

e. 2963.58 _____

• *Using the constant function*

2-4

Who's a Cube?

Find out about squares and cubes

Intermediate Calculator Math

Squaring

Multiplying a number by itself (say, 6 × 6) is called "squaring" the number.
Another way of writing 6 × 6 is 6^2 (read "six squared").

Find: $6^2 =$ _____ $9^2 =$ _____ $8^2 =$ _____ $41^2 =$ _____ $21^2 =$ _____

$3^2 =$ _____ $4^2 =$ _____ $22^2 =$ _____ $93^2 =$ _____ $88^2 =$ _____

A fast method of squaring a number such as 6 on your calculator is by pressing the numeral
key $\boxed{6}$, then the multiplication button $\boxed{\times}$ and then the equals sign $\boxed{=}$.
The answer 36 will appear.
Check all of your answers for the questions above, using the new method.

Cubing

Multiplying a number by itself twice (say, 4 × 4 × 4) is called "cubing" the number.
Another way of writing 4 × 4 × 4 is 4^3 (read "four cubed").

Find: $4^3 =$ _____ $9^3 =$ _____ $2^3 =$ _____ $17^3 =$ _____ $97^3 =$ _____

$3^3 =$ _____ $7^3 =$ _____ $8^3 =$ _____ $31^3 =$ _____ $100^3 =$ _____

Using the information from "Squaring," can you find a fast method for cubing,
using your calculator?

Fill in the table. Some answers are given as hints.

Copyright © 1980

"Pushes"						
7						
6						
5					$4^5=1024$	
4						
3			$2^3 = 8$			
2						
Start	1	1	2	3	4	5

• *Learning about exponents*
• *Optimizing the use of the calculator*

Name _____

2-5

What Multiplication Is: Part I

Look at the pictures two ways

1.

$4 + 4 + 4 =$ _____

$3 \times 4 =$ _____

2.

$5 + 5 =$ _____

$2 \times 5 =$ _____

3.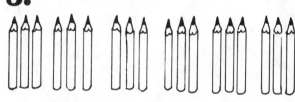

$3 + 3 + 3 + 3 + 3 + 3 =$ _____

$6 \times 3 \quad\quad =$ _____

4.

$5 + 5 + 5 + 5 + 5 + 5 + 5 =$ _____

$7 \times 5 \quad\quad =$ _____

5.

$8 + 8 + 8 =$ _____

____ $\times 8 =$ _____

6.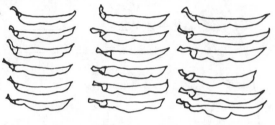

___ + ___ + ___ = _____

___ × ___ = _____

7.

___ + ___ + ___ + ___ + ___ = _____

___ × ___ = _____

8.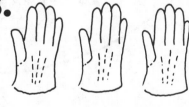

___ + ___ + ___ + ___ = _____

___ × ___ = _____

• *Understanding multiplication*
• *Optimizing the use of the calculator*

Name _____ 15

2-6

What Multiplication Is: Part II

Which way is faster?

Intermediate Calculator Math

1. $8 + 8 + 8 + 8 = $ _____

$4 \times 8 \qquad = $ _____

2. $2 + 2 + 2 = $ _____

___ $\times$ ___ $= $ _____

3. $9 + 9 = $ _____

$2 \times 9 = $ _____

4. $3 + 3 + 3 + 3 + 3 + 3 + 3 + 3 = $ _____

$8 \times 3 \qquad\qquad = $ _____

5. $7 + 7 + 7 + 7 + 7 = $ _____

___ $\times 7 \qquad = $ _____

6. $6 + 6 + 6 + 6 + 6 = $ _____

$5 \times 6 \qquad = $ _____

7. $4 + 4 + 4 + 4 + 4 = $ _____

$5 \times$ ___ $\qquad = $ _____

• *Understanding multiplication*
• *Optimizing the use of the calculator*

Name _____

2-7

The Commutative Property

How addition and multiplication are related

Complete. (Check your work with a calculator.)

1. $5 \times 4 = 4 + 4 + 4 + 4 + 4 =$ _____
$4 \times 5 = 5 + 5 + 5 + 5 =$ _____

2. $9 \times 3 = 3 + 3 + 3 + 3 + 3 + 3 + 3 + 3 + 3 =$ _____
$3 \times 9 = 9 + 9 + 9 =$ _____

3. $6 \times 5 = 5 + 5 + 5 + 5 + 5 + 5 =$ _____
$5 \times 6 = 6 + 6 + 6 + 6 + 6 =$ _____

4. $8 \times 3 = 3 + 3 + 3 + 3 + 3 + 3 + 3 + 3 =$ _____
$3 \times 8 = 8 + 8 + 8 =$ _____

5. $5 \times 2 = 2 + 2 +$ _____ = _____
$2 \times 5 =$ _____ = _____

6. $7 \times 6 =$ _____ = _____
$6 \times 7 =$ _____ = _____

7. $9 \times 4 =$ _____ = _____
$4 \times 9 =$ _____ = _____

8. $8 \times 5 =$ _____ = _____
$5 \times 8 =$ _____ = _____

Illustrate your answers with dot diagrams.

• *Reinforcing the concept of the commutative property*

Name _____

2-8

The Important Order

What you put first makes a difference

a.

$4 + 5 \times 3 = 19$		$6 + 7 \times 4 = 34$		$2 + 5 \times 4 = 22$	
Agree	Disagree	Agree	Disagree	Agree	Disagree

b.

$5 \times 3 + 4 =$	$7 \times 4 + 6 =$	$5 \times 4 + 2 =$
Correct answer: 19	Correct answer: 34	Correct answer: 22

c.

$6 + 4 \times 5 =$	$9 + 3 \times 5 =$	$5 \times 8 + 2 =$
Rewritten: (if necessary) _____ Answer: _____	Rewritten: (if necessary) _____ Answer: _____	Rewritten: (if necessary) _____ Answer: _____
Correct answer: 26	Correct answer: 24	Correct answer: 42

$6 + 7 \times 4 =$	$8 \times 3 + 7 =$	$8 \times 7 + 4 =$
Rewritten: (if necessary) _____ Answer: _____	Rewritten: (if necessary) _____ Answer: _____	Rewritten: (if necessary) _____ Answer: _____
Correct answer: 34	Correct answer: 31	Correct answer: 60

d.

1. $2 + 7 \times 5 =$ _____   **3.** $9 \times 5 + 4 =$ _____

2. $8 \times 6 + 2 =$ _____   **4.** $8 + 7 \times 6 =$ _____

- *Discovering the need for operation-order rules*
- *Optimizing the use of the calculator*

Intermediate Calculator Math

Copyright © 1980

Name _____

2-9

Careful Comparisons

Just by looking, can you tell?

Insert <, =, or > to make the statements true.
Use your calculator to check.
The first problem is done for you.

Note

348 + 620   <   350 + 620

244 + 127   240 + 130

389 + 538   390 + 540

449 + 860   450 + 860

149 + 296   150 + 300

681 + 766   680 + 770

868 + 873   870 + 870

2801 + 1001   2800 + 1000

2945 + 9584   2950 + 9580

3595 + 5715   3600 + 5710

3087 + 9499   3090 + 9500

4934 + 1536   4930 + 1540

Intermediate Calculator Math

• *Estimating in addition*
• *Seeing number relations*

2-10

Triple Choice

A real test of your estimating power

Intermediate Calculator Math

Select the smallest number, the middle number, and the largest number on each line.
Mark them A, B, and C respectively.
The first one is done for you.

1.	**C** 698 + 544	**B** 328 + 443	**A** 107 + 343
2.	___ 284 + 625	___ 595 + 237	___ 304 + 990
3.	___ 487 + 964	___ 107 + 748	___ 432 + 622
4.	___ 128 + 969	___ 195 + 164	___ 642 + 116
5.	___ 340 + 865	___ 354 + 307	___ 283 + 505
6.	___ 727 + 909	___ 642 + 111	___ 307 + 119
7.	___ 238 + 109	___ 116 + 369	___ 430 + 277
8.	___ 629 + 324	___ 210 + 897	___ 521 + 206
9.	___ 435 + 802	___ 382 + 208	___ 705 + 284
10.	___ 732 + 239	___ 692 + 677	___ 257 + 631
11.	___ 564 + 916	___ 125 + 533	___ 616 + 212
12.	___ 427 + 424	___ 227 + 505	___ 784 + 611

Now check with your calculator.

• *Estimating three-digit numbers*
• *Rounding to the nearest ten or hundred*

Name _____

2-11

The Educated Guess

How close can you come how fast?

Record your guess for the sum of the numbers in each circle.
Then check with your calculator.

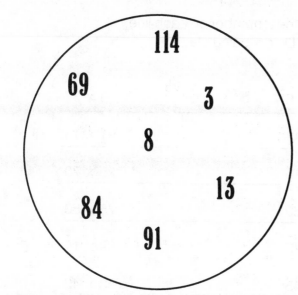

1. Guess: _____ Exact sum: _____
Was your guess too high or too low?

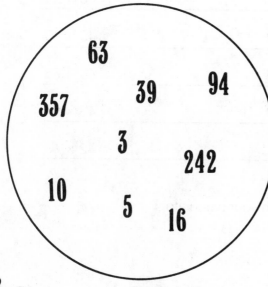

3. Guess: _____ Exact sum: _____
Was your guess too high or too low?

2. Guess: _____ Exact sum: _____
Was your estimate too high or too low?

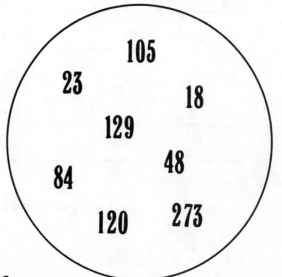

4. Estimate: _____ Exact sum: _____
Was your estimate too high or too low?

• *Estimating in addition*

Name _____

2-12

Tri-Option

Biggest number gets a circle

Use rounding to select the largest number on each line.
Circle your answer. Do not calculate.

53 × 7	50 × 8	19 × 37
38 × 29	23 × 42	91 × 13
47 × 34	90 × 16	25 × 70
89 × 41	60 × 60	75 × 50
25 × 49	75 × 20	89 × 15
23 × 94	45 × 45	53 × 42
47 × 61	21 × 94	53 × 57
54 × 54	49 × 74	68 × 29
84 × 43	62 × 63	45 × 90
56 × 72	65 × 65	45 × 85
61 × 70	29 × 99	43 × 97
44 × 39	89 × 17	42 × 48

Check your answers with your calculator.

Intermediate Calculator Math

Copyright © 1980

• *Estimating in multiplication*
• *Rounding to the nearest ten*

Name _____

2-13

Home, Home on the Range: Part I

You choose the challenge

a. Write your estimate for the range into which the answer will fall.
Make the range as small as you dare.

b. Check by finding the exact answer with your calculator.

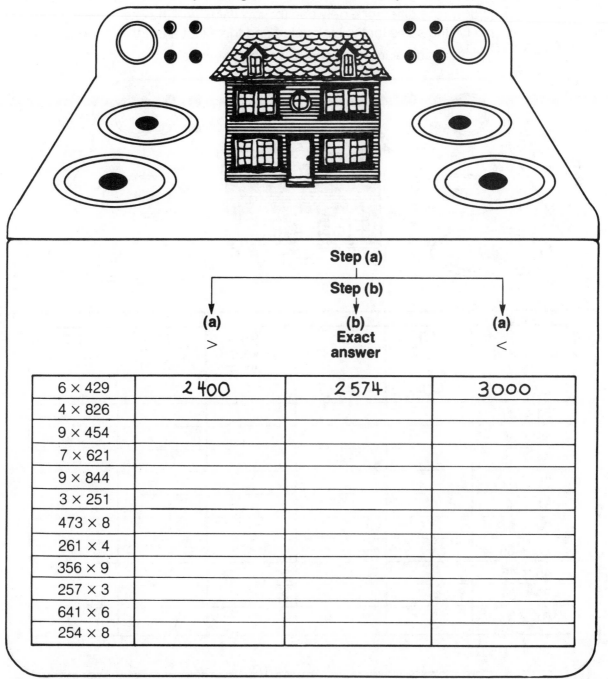

	Step (a) → (a) >	Step (b) ↓ (b) Exact answer	Step (a) → (a) <
6 × 429	2400	2574	3000
4 × 826			
9 × 454			
7 × 621			
9 × 844			
3 × 251			
473 × 8			
261 × 4			
356 × 9			
257 × 3			
641 × 6			
254 × 8			

• *Estimating in multiplication*
• *Developing the concept of a range*

Intermediate Calculator Math

Name _____

2-14

Home, Home on the Range: Part II

Check your estimating habits

a. Write your estimate for the high and low values between which the answer will fall. Make the difference between the high and low as small as you dare.

b. Check by finding the exact answer with your calculator.

c. Write down the difference between the low value and the exact answer. Do the same for the high value and the exact answer.

d. Add columns (d) and (e) to find your total score.

	(a) >	(b) <	(c) Exact answer	(d) |c − a|	(e) |b − c|
23 × 19	400	460	437	37	23
51 × 37					
49 × 36					
54 × 48					
96 × 42					
51 × 36					
94 × 94					
83 × 68					
45 × 45					
73 × 81					
61 × 93					

Score:

• *Estimating in multiplication*
• *Using the concept of a range*
• *Judging error in the estimation process*

Name _____

Intermediate Calculator Math

2-15

Home, Home on the Range: Part III

How well do you estimate bigger numbers?

a. Write your estimate for the range into which the answer will fall.
Make the range as small as you dare.

b. Check by finding the exact answer with your calculator.

c. Write down the difference between the low of the range and the exact answer.
Do the same for the high of the range and the exact answer.

d. Add columns (d) and (e) to find your score.

	(a) >	(b) <	(c) Exact answer	(d) |c − a|	(e) |b − c|
60 × 437	25,500	27,000	26,220	720	780
40 × 965					
70 × 259					
80 × 496					
70 × 654					
137 × 90					
446 × 20					
50 × 475					
625 × 40					
343 × 20					
545 × 60					

Score:

• *Estimating in multiplication*
• *Using the concept of a range*
• *Judging error in the estimation process*

Intermediate Calculator Math

Name _____

2-16

Home, Home on the Range: Part IV

Try to get these even closer!

a. Write your estimate for the range into which the answer will fall.
Make the range as narrow as you can.
b. Check by finding the exact number with your calculator.
c. Write the difference between the low of the range and the exact answer.
Do the same for the high of the range and the exact answer.
d. Add columns (d) and (e) to find your score.

	(a) >	(b) <	(c) Exact answer	(d) |c − a|	(e) |b − c|
300 × 64	18,000	19,500	19,200	1200	300
700 × 85					
900 × 61					
600 × 43					
700 × 29					
49 × 500					
62 × 700					
91 × 600					
37 × 400					
73 × 800					
34 × 900					

Score:

Intermediate Calculator Math

Copyright © 1980

• *Estimating in multiplication*
• *Using the concept of a range*
• *Judging error in the estimation process*

Name _____

2-17

Home, Home on the Range: Part V

The numbers are larger but don't give up

a. Write your estimate for the range into which the answer will fall.
Make the range as narrow as you dare.
b. Check by finding the exact answer with your calculator.
c. Write down the difference between the low of the range and the exact answer.
Do the same for the high of the range and the exact answer.
d. Add columns (d) and (e) to find your score.

	(a) >	(b) <	(c) Exact answer	(d) $|c - a|$	(e) $|b - c|$
174 × 61	10,000	11,000	10,614	614	386
37 × 451					
26 × 934					
97 × 614					
298 × 47					
361 × 94					
369 × 247					
451 × 361					
258 × 414					
309 × 646					
747 × 818					

Score:

Intermediate Calculator Math

Copyright © 1980

• *Estimating large products*
• *Using the concept of a range*
• *Judging error in the estimation process*

Name _____ 27

2-18

Home, Home on the Range: Part VI

Try for an even lower score!

a. Write your estimate for the low and high values between which the answer will fall. Make the difference between these values as small as you can.

b. Check by finding the exact answer with your calculator.

c. Write the difference between your low estimate and the exact value. Do the same for your high estimate.

d. Add columns (d) and (e) to find your score.

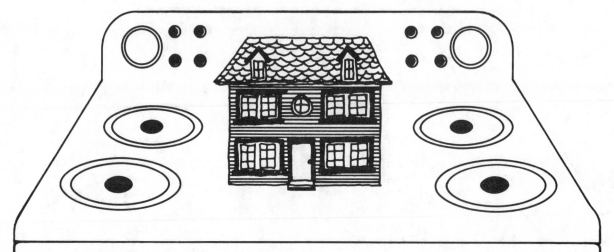

Intermediate Calculator Math

	(a) >	(b) <	(c) Exact answer	(d) |c − a|	(e) |b − c|
7 × 61					
30 × 84					
97 × 40					
385 × 174					
80 × 672					
37 × 42					
800 × 258					
474 × 800					
60 × 493					
300 × 296					
897 × 6					

Score:

- *Estimating large products*
- *Using the concept of a range*
- *Judging error in the estimation process*

Name _____

2-19

Somewhere In-between

Several answers can be equally right!

Fill in the blanks to get an answer in the given range.
The first one is done for you.

Multiplication	Range
34 × __2.5__	70 ←→ 90
56 × _____	300 ←→ 350
23 × _____	600 ←→ 650
17 × _____	500 ←→ 550
98 × _____	450 ←→ 500
12 × _____	100 ←→ 120
22 × _____	480 ←→ 520
38 × _____	220 ←→ 280
64 × _____	950 ←→ 1000
31 × _____	600 ←→ 650
75 × _____	650 ←→ 700
52 × _____	820 ←→ 880
49 × _____	840 ←→ 880
61 × _____	200 ←→ 240
26 × _____	140 ←→ 170
33 × _____	830 ←→ 870
41 × _____	570 ←→ 610
24 × _____	860 ←→ 900
57 × _____	250 ←→ 280
19 × _____	475 ←→ 500

• *Estimating in multiplication*
• *Estimating in division*

Intermediate Calculator Math

Name _____

2-20

A Santa Claus Riddle

What advice does Santa Claus give to farmers?

Answer

Color in the division expression with the biggest answer for each row (a, b, c, etc.).

In rows c, e, and g, color in the 2 largest expressions for the row.

Estimate, don't work them out!

You may check with your calculator.

Row "a" is done for you.

Rotate the page to find the answer!

Intermediate Calculator Math

Copyright © 1980

a	34	÷	5		69	÷	7		167	÷	9
b	45	÷	7		129	÷	12		12	÷	3
c	555	÷	25		26	÷	8		29	÷	4
d	67	÷	6		337	÷	19		37	÷	8
e	57	÷	1		34	÷	3		918	÷	11
f	118	÷	8		58	÷	2		465	÷	64
g	834	÷	92		52	÷	9		237	÷	6
h	50	÷	7		91	÷	2		831	÷	36
i	92	÷	3		228	÷	17		16	÷	3

• *Estimating in division*

Name _____

2-21

The Guess and Check Game

A division-facts challenge   (for 2 players)

Equipment
Calculator
Score sheets (below)
Pencil

Instructions

1. Player A punches any 3-digit number (say, 547) into the calculator.

2. Player B punches ⌑÷⌑, followed by a 2-digit number (say, 27).

3. Both players write the question, and their estimates of the answer, on their score sheets.

4. Player A pushes the ⌑=⌑ button to determine the actual answer.
   This is recorded, along with the difference between their guesses and this answer.

5. The player with the lower difference wins the round.

6. A "best-of-10" series is played.

Player A	**Name:**		
Question	**Estimate**	**Actual**	**Diff.**
÷			
÷			
÷			
÷			
÷			
÷			
÷			
÷			
÷			
÷			

Player B	**Name:**		
Question	**Estimate**	**Actual**	**Diff.**
÷			
÷			
÷			
÷			
÷			
÷			
÷			
÷			
÷			
÷			

• *Estimating in division*
• *Refining estimation abilities for accuracy*

Name_____ 31

2-22

The Covered Target Game

So you know your addition facts?

(for 2 players)

Intermediate Calculator Math

Equipment

Calculator (display covered)
Score sheet
Pencil

Instructions

1. Players agree upon a "target" number (say, 60).

2. While player B watches, player A punches in a single-digit number (say, 6). Player B then adds a single-digit number (say, 7). Players then take turns adding single-digit numbers. ("0" is not allowed!)

3. Instead of taking a turn, a player who feels that the display (hidden) is as close to the target as it will get may call "Stop". The display is examined and the caller is awarded the difference between the display and the target. (for display: 58, target: 60, score is 2.)

4. This is a 10-round game. The game will work best if you imagine that in round 1 you called "stop" and scored 10, while in round 2 your partner called "stop" and scored 10. Play the remaining 8 rounds to see who can get the *lowest average* score (if you call "stop" 4 times and your scores are 10, 7, 3, 8, your average is $(10 + 7 + 3 + 8) \div 4 = 7$).

Round	Target	Player A	Player B
1	60	10	
2	48		10
3			
4			
5			
6			
7			
8			
9			
10			
Total		_____	_____
Average		_____	_____

(Tape over display)

FLAP

- - - - - - - - - - - - -

ANSWER BELOW

Don't lift until "Stop" is called!

- *Reinforcing addition facts*
- *Averaging*

Name _____

2-23

Monstrous Multiplication

Search for the right pattern

(for 2 to 6 players)

Equipment
20 mini-cards (below) cut out, shuffled, and placed in a box
Calculator
Pencil and paper

Instructions
Players take turns selecting cards and placing them face up on the table, until a total of 6 cards is revealed. Each of the players, in turn, arranges the digits into 2 numerals so that the *product* will be as large as possible. Write down the numerals. Use the calculator to compare results. The player who has the largest product wins.

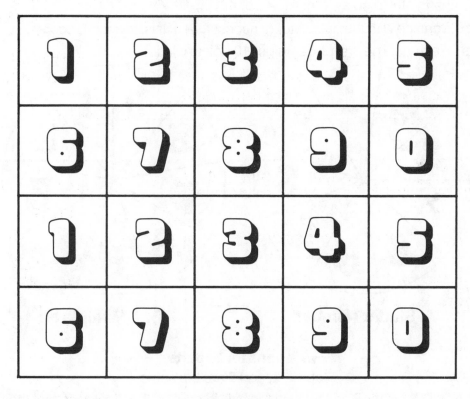

Variations
a. The player with the smallest product wins.
b. Increase the number of cards drawn to 8 or 10.

Note: You may discover a general algorithm for this game.

Intermediate Calculator Math

• *Estimating in multiplication*
• *Detecting patterns*

Name _____ 33

2-24

Divide and Conquer

A game of factors and multiples   (for 2 players)

Intermediate Calculator Math

Equipment
Spinners (cut from this page)
2 thumb tacks   3 pencils   Calculator

Instructions

1. Complete the 2 spinners on this page by pushing a thumbtack through their centers from the back and sticking a pencil on the point of the tack to act as a dial.

2. Player A spins spinner A and displays the result on the calculator (say, 4).

3. Player B then flicks the other spinner to determine the "factor" (say, 6). B must now place one more digit (his or her choice) on the display to make it a multiple of 6 (B can add 2 or 8 to change the display from 4 to 42 or 48).

4. Check B's choice with the calculator (if necessary). If it is correct, B scores a point.

5. Players take turns. The first player to get 10 points wins.

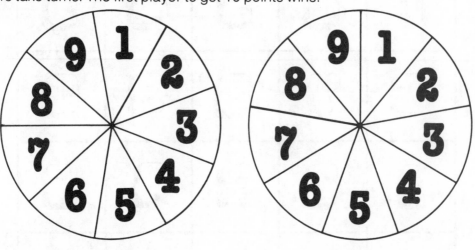

A. Starter Number   **B. A Multiple of:**

Variations for Experts
Steps **1, 2,** and **3** are the same.

4. Player A flicks spinner B and must place a third digit on the display so that the number displayed is a multiple of the number on the spinner.

5. If the number displayed is already a multiple of the number on the spinner, Player A has the option of saying "Pass."

6. The first player forced beyond the display loses, unless someone has displayed an incorrect multiple first.

Copyright © 1980

• *Reviewing multiplication and division*
• *Factoring and multiples*

Name _____

2-25

Roundabout Route to a Square Root

A memory-and-estimation-challenge   (for 2 players)

Equipment
Calculators
Set of cards (below)

Instructions

1. Cut out the cards, shuffle them, and place them face down in a pile.

2. Player A turns over the top card. He or she estimates the square root of the number on the card, and checks with the calculator (b × b = ?).

3. Player B may watch A's attempt. If A's guess is wrong, B takes a guess. Players continue, taking turns "narrowing in" on the target.

4. The player who finds the correct square root keeps the card.

5. Play continues until all cards have been used. The player with the most cards wins.

2704	309,136	734,449	6084
7921	113,569	169,744	1936
6561	4624	620,944	341,056
1296	1849	603,729	622,521
66,564	3136	9409	15,376

• *Estimating in multiplication or division*
• *Using square roots*

Intermediate Calculator Math

Name _____

2-26

Destination 500

Getting there is half the fun

(for 2 to 4 players)

Intermediate Calculator Math

Equipment
Calculator
2 score sheets, pencils

Instructions

1. Player A selects a number (say, 500) as the destination. Both players write it on their score sheets.

2. Player B announces the starting number (say, 13). Both players record this number, also.

3. Each player now writes down a "multiplier" (a number to multiply by 13 to get 500). Decimals are permitted.

4. The player whose product is closer to 500 wins the round. (Check with the calculator.) The winner puts a check mark in the ✔ column of his or her score sheet.

5. After each round, a new destination and a new starting number are chosen.

6. The player who wins the most rounds out of 10 is the champion.

Player A: _____					Player B: _____				
Destination	**Starting Number**	**Multiplier**	**Product**	✔	**Destination**	**Starting Number**	**Multiplier**	**Product**	✔
a					a				
b					b				
c					c				
d					d				
e					e				
f					f				
g					g				
h					h				
i					i				
j					j				

• *Reviewing multiplication facts*
• *Estimating in multiplication and division*

Copyright © 1980

Name _____

2-27

Two Astounding Tricks

Learn these and try on your friends

Equipment
Calculator
Pencil
Paper

1. Chain Trick   **Examples**

a. Pick 3 one-digit numbers. Any 3 work!   4, 6 and 2

b. Multiply the first number by 2. $2 \times 4 = 8$

c. Add 3. $8 + 3 = 11$

d. Multiply by 5 and add 7. $11 \times 5 + 7 = 55 + 7 = 62$

e. Add the second number. $62 + 6 = 68$

f. Multiply by 2 and add 3. $68 \times 2 + 3 = 136 + 3 + 139$

g. Multiply by 5 and add the third number. $139 \times 5 + 2 = 695 + 2 = 697$

h. Deduct 235. $697 - 235 = 462$

The result will be the 3 digits in the order they were used!

2. Chain Trick

a. Pick any 2 *consecutive* numbers.   5000, 5001

b. Add them together. $5000 + 5001 = 10,001$

c. Add nine. $10,001 + 9 = 10,010$

d. Divide by 2. $10,010 \div 2 = 5005$

e. Subtract the smaller of the two numbers you started with. $5005 - 5000 = 5$

Your answer is _____.

Repeat these steps using several different numbers and record your answers.
Amazing???
What is *your* explanation of this result?

• *Developing algebraic reasoning*

*Name*_____

2-28

Faster Than a Speeding Calculator

You can beat the machine!   (for 2 players)

Intermediate Calculator Math

Equipment
Calculator
Pencil and paper

Instructions

Example 1

a. Have your partner write a 3-digit number.   527

b. Have your partner write another 3-digit number.   231

c. You write 768. (Think 231 + ____ = 999.)   768

d. Tell your partner to add, using the calculator, then you give the answer before   1526
your partner can finish!

(The numbers in the boxes are related. How?)

Example 2

a. Have your partner write a 4-digit number.   4253

b. Have your partner write another 4-digit number.   5623

c. You write 4376. (Think 5623 + _____ = 9999.)   4376

d. Have your partner write another 4-digit number.   2341

e. You write 7658. (Think 2341 + _____ = 9999.)   7658

f. Give the answer before your partner can find it using the calculator!   24251

(The numbers in the boxes are related. How?)

Note: Once you've mastered this, you may want to try it with 5-digit numbers.

Copyright © 1980

• *Developing algebraic reasoning*
• *Discovering number relationships*

Name _____

2-29

A Fraction by Another Name

What can you call these spaces?

Assign a common fraction to the shaded part in each drawing.
Then use your calculator to find the corresponding decimal fraction.

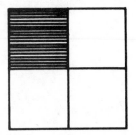

a. _____ = _____

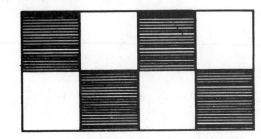

b. _____ = _____

c. _____ = _____

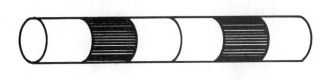

d. _____ = _____

0 ├———┼———┼———┼——▭——┤ 1

f. _____ = _____

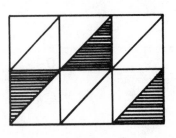

e. _____ = _____

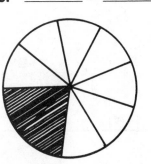

g. _____ = _____

h. _____ = _____

• *Understanding the concept of fractions*
• *Converting fractions to decimal form*

2-30

The Fraction Line

How do you tell where they fall?

Place these points on the number line.
The first one is done for you.

1. A = $\frac{3}{5}$ = 0.6   B = $\frac{4}{9}$ C = $\frac{4}{7}$ D = $\frac{3}{6}$

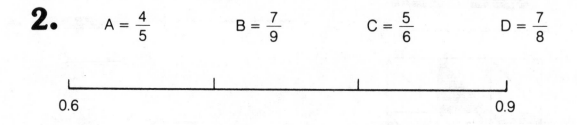

2. A = $\frac{4}{5}$ B = $\frac{7}{9}$ C = $\frac{5}{6}$ D = $\frac{7}{8}$

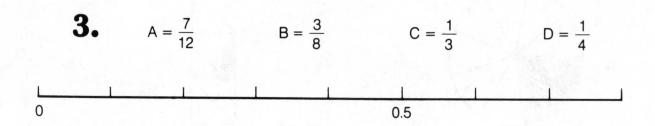

3. A = $\frac{7}{12}$ B = $\frac{3}{8}$ C = $\frac{1}{3}$ D = $\frac{1}{4}$

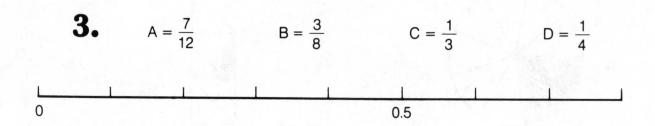

• *Converting fractions to decimal form*
• *Comparing fractions*

Name _____

Intermediate Calculator Math

2-31

Larger Fractions, Smaller Fractions

A matter of comparing

Put the correct sign (>, <, or =) between each pair of fractions.
Check by converting to decimals, using your calculator.

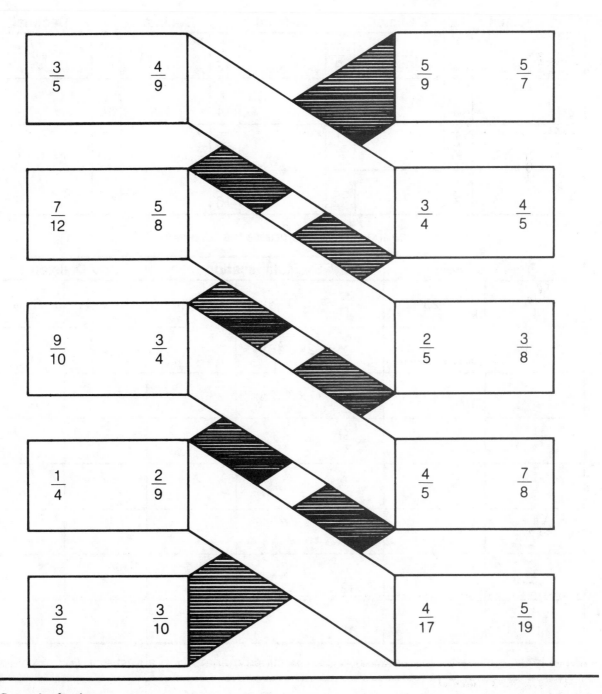

• **Comparing fractions**
• **Converting fractions to decimal form**

2-32

A Fraction by Another Name: II

Each fraction has a decimal "alias"

Find the decimal name for each number.
Carry out the division to as many decimal places as your calculator allows.

$\frac{1}{1}$	Decimal	$\frac{1}{5}$	Decimal	$\frac{1}{9}$	Decimal	$\frac{1}{13}$	Decimal	$\frac{1}{17}$	Decimal
$\frac{1}{2}$		$\frac{1}{6}$		$\frac{1}{10}$		$\frac{1}{14}$		$\frac{1}{18}$	
$\frac{1}{3}$		$\frac{1}{7}$		$\frac{1}{11}$		$\frac{1}{15}$		$\frac{1}{19}$	
$\frac{1}{4}$		$\frac{1}{8}$		$\frac{1}{12}$		$\frac{1}{16}$		$\frac{1}{20}$	

Use the table to complete the following.

From the Table		By Multiplication		By Division		
$\frac{1}{4}$	0.25	$\frac{3}{4} = (3 \times 0.25)$	0.75	$3 \div 4$	0.75	
$\frac{1}{6}$	$0.1\overline{6}$	$\frac{2}{6} = (2 \times 0.1\overline{6})$		$2 \div 6$		*
$\frac{1}{8}$	0.125	$\frac{7}{8} = (7 \times 0.125)$		$7 \div 8$		
$\frac{1}{3}$		$\frac{2}{3}$		$2 \div 3$		
$\frac{1}{9}$		$\frac{5}{9}$				
$\frac{1}{10}$		$\frac{4}{10}$				
$\frac{1}{12}$		$\frac{7}{12}$				*
$\frac{1}{18}$		$\frac{11}{18}$				*

What is unusual about the answers to the questions marked with an asterisk (*)? Explain.

- *Converting fractions to decimal form*
- *Building a table*
- *Comparing processes for determining decimal equivalents*

Name _____

2-33

Guess How It Goes From Here

Look closely at what repeats

1. Use your calculator to find the decimal equivalent for the first three fractions in the group. Then guess the others.

Calculate: Guess:

$\dfrac{1}{3}$ = _____ $\dfrac{1}{3333}$ = _____

$\dfrac{1}{33}$ = _____ $\dfrac{1}{33,333}$ = _____

$\dfrac{1}{333}$ = _____   Check with your calculator.

2.

Calculate: Guess:

1×1 = _____ 1111×1111 = _____

11×11 = _____

111×111 = _____   Check with your calculator.

These products go beyond the calculator display. Predict the answers. How can they be checked?

$11,111 \times 11,111 =$ _____ $111,111 \times 111,111 =$ _____

3.

Calculate: Guess:

9×9 = _____ 9999×9999 = _____

99×99 = _____

999×999 = _____   Check with your calculator.

These products go beyond the calculator display. Can you predict the answers? How can they be checked?

$99,999 \times 99,999 =$ _____ $999,999 \times 999,999 =$ _____

• *Detecting and completing patterns*
• *Using intuition*

Intermediate Calculator Math

Name _____ 43

2-34

Pattern Discovery

Discovering your own with a number array

Consider the array below.

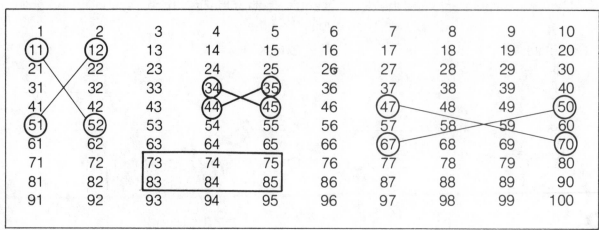

1. Compare the pair 34, 45 with the pair 44, 35.

    a. Add: 34 + 45 = _____

            44 + 35 = _____

    What do you notice about the answers?

    _____

    Is that finding true for any two such pairs?

    _____

    b. Subtract: 45 − 34 = _____

               44 − 35 = _____

    How do the answers differ?

    _____

    Is that true for any two such pairs?

    _____

    c. Multiply: 34 × 45 = _____

              35 × 44 = _____

    How do the answers differ?

    _____

    Is that true for any two such pairs?

2. There are many other patterns you can find.

Describe some of the patterns you found.

Do they hold true if the array is extended to include negative numbers?

3. What happens to your pattern if the array is changed as follows?

1	2	3	4	5	6
7	8	9	10	11	12
13	14	15	16	17	18
19	20	21	22	23	24

Intermediate Calculator Math

• *Detecting and completing patterns*
• *Drawing conclusions*

Name _____

2-35

A Big Fish Story

How much will that fish cost?

Spend $50.
Provide an itemized account of your purchases.

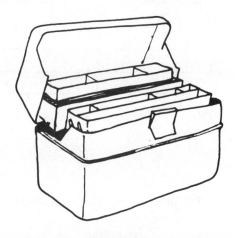

Exclusive!
3 Drawers, Floating, 1-year Guarantee
$14.95

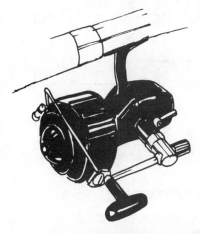

Playmaker Spinning Reel
$10.95

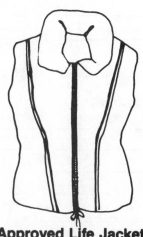

Approved Life Jacket
$12.95

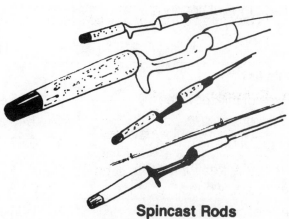

Spincast Rods
With off-set handles for
closed-faced reels
$29.95

• *Estimating in addition*
• *Gathering and organizing data*
• *Solving problems*

Name _____

2-36

Stocking Up

Estimate the costs of this campout

The girl scouts are going camping. Three girls have been assigned to buy the supplies.
Estimate how much money each scout should be given for supplies.
Then use these estimates to estimate the total amount of money needed.
Finally, check all figures with your calculator.

Peggy's list		Estimate	Actual
30 oranges	@ $0.15 each	$ 4.00	
6 heads of lettuce	@ 0.39 each		
4 bags of apples	@ 1.29 each		
5 bunches of carrots	@ 0.49 each		
3 watermelons	@ 1.98 each		
2 baskets of tomatoes	@ 1.49 each		
Total			

Kathleen's list		Estimate	Actual
12 cans of beans	@ $0.32 each	$ 3.50	
4 cans of tomato soup	@ 0.26 each		
6 cans of mushroom soup	@ 0.29 each		
4 jars of peanut butter	@ 1.24 each		
2 jars of strawberry jam	@ 1.29 each		
4 packages of hot dogs	@ 1.12 each		
Total			

Patti's list		Estimate	Actual
2 boxes of milk powder	@ $2.25 each		
2 boxes of soap	@ 1.49 each		
6 loaves of bread	@ 0.47 each		
3 boxes of breakfast cereal	@ 1.12 each		
4 packages of buns	@ 0.59 each		
2 bags of cookies	@ 1.38 each		
Total			

Estimated total sum of money needed for the 3 girls: _____

Actual total sum of money needed for the 3 girls: _____

• *Estimating in multiplication*
• *Building consumer math skills*

Name _____

Intermediate Calculator Math

2-37

An Expensive House!

See how the materials add up

Mary and her parents are building a playhouse.
As you can see, they have already purchased many of the materials.
Estimate the total cost of the materials so far.

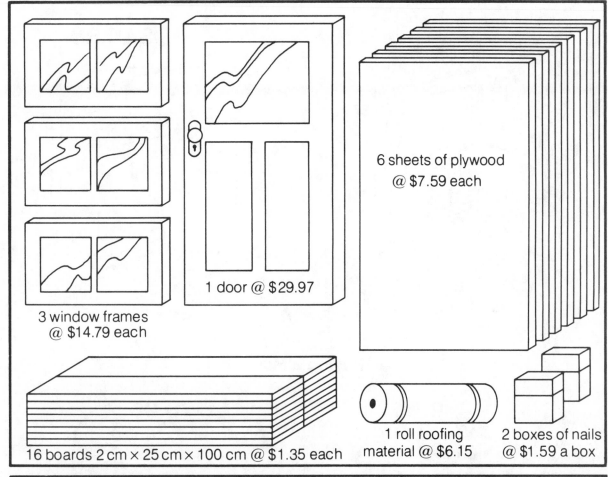

6 sheets of plywood
@ $7.59 each

1 door @ $29.97

3 window frames
@ $14.79 each

16 boards 2 cm × 25 cm × 100 cm @ $1.35 each

1 roll roofing
material @ $6.15

2 boxes of nails
@ $1.59 a box

Item	Estimate	Actual	Item	Estimate	Actual
Roofing			Plywood		
Windows			Nails		
Boards			Door		

Estimated Total _____

Actual Total _____

Use your calculator to check your answers.

• *Estimating in multiplication*
• *Gathering information*
• *Building consumer math skills*

Intermediate Calculator Math

Name _____

2-38

Finding the Best Buy

Think first before you spend

Intermediate Calculator Math

Copyright © 1980

Buying a Freezer

0.78 m **$339.98**

1.87 m

0.82 m

$349.98

0.82 m

1.08 m

1.93 m

Which freezer is the best buy?
Why?

Buying a Mower

$89.95

Width of cut: 50 cm

$79.95

Width of cut: 45 cm

Which mower is the best buy?
Why?

• *Working with volume and ratio*
• *Calculating cost per unit*

Name _____

2-39

The World's Largest Swimming Pool

Figure areas and volumes

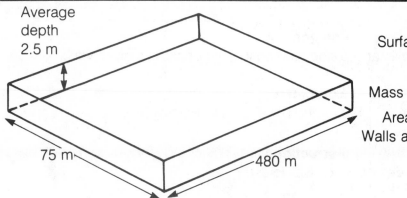

Average depth 2.5 m

75 m   480 m

Surface Area = _____

Volume = _____

Mass of Water = _____

Area of Pool
Walls and Floor = _____

_____

0.42 m

1.97 m

1.07 m

1.96 m

1.98 m

2.44 m   3.45 m

Volume = _____

Surface Area

(with floor) = _____

(without floor) = _____

Royal Regency
round model

0.97 m

3.16 m

Capacity _____ L

Copyright © 1980

Intermediate Calculator Math

• *Working with area and volume*
• *Gathering information from drawings*
• *Working with pi*

2-40

Brick Trick

A real-world problem

(for 2 students)

Intermediate Calculator Math

Assignment

to determine the approximate number and cost of the bricks in a given wall.

Equipment

Calculator   Paper and pencil   Meter stick

Instructions

1. Gather your equipment and decide which wall to use.

2. Measure off a square 1 m by 1 m on the brick wall. Its area is 1 m².

3. Carefully count the bricks within the square. Decide with your partner what to do about the half bricks or other part bricks that fall within the square. Record the number on this sheet. Number of bricks in 1 m²: _____

4. First estimate, then measure or calculate, the length and height of the wall.

Estimated length of wall: _____

Estimated height of wall: _____

Measured length of wall: _____

Measured or calculated height of wall: _____

5. Estimate, then calculate, the total number of bricks on the surface of the wall.

Estimated number of bricks: _____ Calculated number of bricks: _____

6. Ask your teacher the cost of 1 brick.

Cost of 1 brick: _____

7. Estimate, then calculate, the cost of the bricks on the surface of the wall.

Estimated cost: _____   Actual cost: _____

8. Design a brick clubhouse. Draw the plan on the back of this sheet. Include the main measurements. Estimate, then calculate, the number and cost of the bricks to build it.

Estimated number of bricks: _____ Calculated number of bricks: _____

Estimated cost of bricks: _____ Calculated cost of bricks: _____

Copyright © 1980

• *Gathering data*
• *Constructing diagrams*
• *Working with area*

Name _____

2-41

A Drop in the Bucket: Part I

Collect your own data first

(with a friend)

Copyright © 1980

Intermediate Calculator Math

Equipment
Measuring cup or graduated cylinder
Container for collecting water
Stop watch or clock with second hand
Calculator

Instructions

1. Collect all materials.

2. Find a water tap that no one else will need during the next half hour.

3. Set the tap dripping slowly.

4. Once the tap is dripping steadily, have one person put the container under the tap to collect the water. Meanwhile the other student observes and writes down on this sheet the exact time, to the second, that the collection started.
Starting time: _____

5. The water is to be collected for exactly 24 min. Calculate to the second the exact moment the container is to be removed from underneath the tap. Write on this sheet the time the collection will be finished.
Completing time: _____

6. While the tap is dripping into the container, carefully count the drops for 4 min. One person counts while the other watches the time.

7. Calculate the average number of drops per minute.
Number of drops per minute: _____

8. When the 24 min. are over, remove the container from under the tap. Measure the total volume and compute the volume of each drop. (You will need the information from part 7).
Volume of water after 24 min: _____
Volume of water after 1 min: _____
Volume of 1 drop of water: _____

9. Return all materials except the calculator.

• *Gathering data*
• *Working with volume calculations*
• *Using sampling and averages*
• *Estimating in multiplication*

Name _____

2-42

A Drop in the Bucket: Part II

Use your data to solve problems

10. Estimate the volume of water that would be wasted by each dripping faucet during a 24 h period and during a whole year. Then calculate.

   Estimate of water wasted in a day: _____

   Estimate of water wasted in a year: _____

   Calculated volume of water wasted in a day: _____

   Calculated volume of water wasted in a year: _____

11. This backyard swimming pool needs filling. Find the volume of the pool and calculate how long it would take to fill the pool from 1 dripping faucet. How many dripping faucets would fill the pool in a day? In a week?

   Volume of swimming pool: _____

   Time to fill pool from 1 faucet: _____

   Number of taps to fill pool in a day: _____

   Number of taps to fill pool in a week: _____

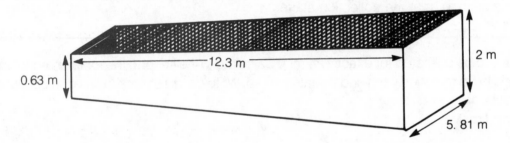

12. Do you think the pool would ever get filled to the top by a single dripping faucet? Why or why not? _____
_____
_____
_____

13. There are about 50,000,000 homes in North America. Assuming each home has 1 dripping faucet, first estimate, then calculate, the total amount of water wasted in all those homes per day and per year.

   Estimate of water wasted per day: _____

   Estimate of water wasted per year: _____

   Calculated volume of water wasted per day: _____

   Calculated volume of water wasted per year: _____

• *Gathering data*
• *Working with volume calculations*
• *Using sampling and averages*
• *Estimating in multiplication*

Name _____

Intermediate Calculator Math

2-43

Superstar Publicity

Figure out the price of fame

Singer I. M. Great publicly promised each of her fans an autographed photo of herself. The response was bigger than expected: all 220,000,000 people in the U.S. wrote for a picture. (Even her own brother!)

Estimate, then calculate, the number of hours I. M. will need to sign all photographs, if she signs 1 picture every 3 s (3 seconds).

Estimated number of hours: _____
Actual number of hours: _____

Estimate, then calculate, the number of 40-h work weeks I. M. will have to spend signing the pictures.

Be careful not to go beyond the capacity of your calculator.

Estimated number of weeks: _____
Calculated number of weeks: _____

I. M. starts signing at age 18. Each year she takes 2 weeks off for vacation. How old will she be when she finishes autographing the photos?

Age: _____   *Danger!*

There are 20 photographs in a thickness of 1 cm. How high is I. M.'s pile of pictures, in kilometers? _____

Each photograph has a mass of 5 g. Estimate, then calculate, the number of large 5 t trucks required to take the photographs to the post office.

Estimated number of 5 t trucks: _____
Actual number of 5 t trucks: _____

2-44

The Point in the Right Place

Size up about what's right

Intermediate Calculator Math

1. Multiply:

46 24	37 68	924 37	65 89	248 37
659 236	514 32	258 467	383 25	296 37
1465 29	1543 614	284 16	495 109	2467 9
454 43	679 28	6794 3144	4645 3175	3882 916

2. Multiply:

4.6 2.4	3.7 68	9.24 3.7	0.65 0.89	0.248 0.37
6.59 2.36	51.4 3.2	0.258 467	383 2.5	2.96 37
146.5 0.29	1543 0.614	0.284 1.6	0.495 1.09	2467 0.09
454 4.3	67.9 28	67.94 3144	4645 3.175	0.3882 916

• *Working with rules for decimal placement*
• *Detecting and explaining patterns*

Name _____

2-45

Solving Sentences: Part I

An educated guess may do it!

Guess which number goes into the box to make the sentence true.
Check with your calculator. If the guess is not quite right,
cross out the answer and try again, and again, until it fits exactly.
An example is done for you.

$4 \times$ | ~~5~~ | ~~5~~ | ~~4.4~~ | 4.5 ✓ | | $+ 5 = 23$

Put your first guess here.

Write your second guess here.

Stop guessing when the answer fits.

1. $2 \times$ [ ][ ][ ][ ][ ] $+ 7 = 16$

2. $5 \times$ [ ][ ][ ][ ][ ] $- 4 = 13$

3. $10 \times$ [ ][ ][ ][ ][ ] $+ 4 = 37$

4. $5 \times$ [ ][ ][ ][ ][ ] $- 13 = 29$

5. $6 \times$ [ ][ ][ ][ ][ ] $+ 14 = 80$

6. $4 \times$ [ ][ ][ ][ ][ ] $- 12 = 14$

7. $10 \times$ [ ][ ][ ][ ][ ] $- 6 = 50$

8. $5 \times$ [ ][ ][ ][ ][ ] $- 3 = 35$

9. $3 \times$ [ ][ ][ ][ ][ ] $+ 5 = 41$

10. $4 \times$ [ ][ ][ ][ ][ ] $- 7 = 28$

11. $10 \times$ [ ][ ][ ][ ][ ] $+ 3 = 48$

12. $4 \times$ [ ][ ][ ][ ][ ] $- 4 = 25$

• *Solving equations with one variable*
• *Applying estimating skills*

Intermediate Calculator Math

Name _____

2-46

Solving Sentences: Part II

Get the answer to this riddle:

Why is a teacher like a candy maker?

clues

1. Find a number (you'll need decimals) that makes each number sentence correct.
2. Write the letter of the problem above the correct answer at the bottom of the page. The first one is done for you.

K: $5 \times \boxed{3.4} + 3 = 20$

T: $6 \times \boxed{} - 9 = 18$

A: $(5 + \boxed{}) \times 5 = 36$

E: $\boxed{}^2 + 0.75 = 7$

U: $\dfrac{3 \times \boxed{}}{2} = 3.9$

H: $4 \times \boxed{} + 3 = 17$

S: $4 \times \boxed{} + 5 = 39$

C: $(2.05 + \boxed{}) \times 8 = 18$

L: $\boxed{}^2 - 0.61 = 9$

L: $5 \times \boxed{} - 3 = 13$

O: $\dfrac{5 \times \boxed{} + 1.5}{4} = 5$

A: $(4.3 + \boxed{}) \times 5 = 23$

answer

They
both

							K				
2.6	8.5	2.5	0.2	3.5	2.2	3.2	3.4	0.3	3.1	3.7	4.5	**!**

- *Solving equations with one variable*
- *Applying estimating skills*

Name _____

Intermediate Calculator Math

2-47

The Pi Discovery

A constant workout

Equipment

Variety of circular objects (wheels, discs, cans, bottle tops, cardboard circles, etc.)
Measuring tapes, rulers
Calculator
Worksheet, pencil

Definitions

Circumference: the distance around the outside of a circle
Diameter: the distance across a circle, passing through the center

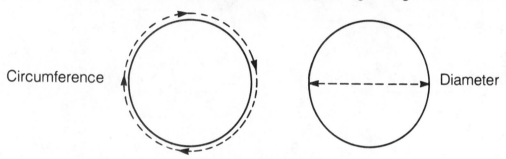

Circumference   Diameter

Instructions

Measure the circumference and diameter of a variety of circular objects.
Write your findings in the table below.
Use your calculator to help you fill in the last column.

Object	Circumference	Diameter	Circumference ÷ Diameter
Wheel	47.8 cm.	15.2 cm	3.144

What conclusion is suggested by the results in the last column?

- *Discovering the relationship between*
  circumference and diameter
- *Gathering data*
- *Putting similar results in tabular form*

Intermediate Calculator Math

Name _____

Answer Key

This section provides solutions for all sheets having exact answers. In estimation exercises, the students' approximations are more important than the exact answers, and the exact answers given here should be de-emphasized. Some sheets have no answers; these are noted in sequence in this section for easy reference.

2-1
The Clever Calculator: Part I
Skip-count fast—or faster

1. Press: 2 + = = = =
Write your answers: 4 6 8 10 12
What's happening? *2 is added to the previous number*

2. Press: 3 + = = = =
Write your answers: 6 9 12 15 18 21
What's happening? *3 is added to each previous total*

3. Press 4 +
What answers will be displayed if you press = = = = = = = Check!
Answers: 8 12 16 20 24

4. Count by 5's, using your calculator. Fill in the blanks.
5, 10, 15, 20, 25, 30, 35, 40, 45, 50, 55, 60, 65, 70, 75, 80

5. Fill in the blanks. Check with your calculator.
6, 12, 18, 24, 30, 36, 42, 48, 54, 60, 66, 72, 78, 84, 90
9, 18, 27, 36, 45, 54, 63, 72, 81, 90, 99, 108, 117, 126, 135
7, 14, 21, 28, 35, 42, 49, 56, 63, 70, 77, 84, 91, 98, 105
20, 40, 60, 80, 100, 120, 140, 160, 180, 200, 220, 240, 260, 280, 300
8, 16, 24, 32, 40, 48, 56, 64, 72, 80, 88, 96, 104, 112, 120
50, 100, 150, 200, 250, 300, 350, 400, 450, 500, 550, 600, 650, 700, 750

• *Introducing the constant feature*
• *Using the calculator optimally*
• *Reviewing multiplication facts*

Name _____

11

2-2
The Clever Calculator: Part II
How the constant helps you

1. a. Clear the calculator and press 2 + three times. (Don't clear.)
Write your answers. 2(4), 4(6), 6(8)
What is happening? *Each time 2 is added to the total.*

c. Clear the calculator and press 2 × 3 Now press = three times. (Don't clear.)
Write your answers. 6, 12, 24
What is happening? *Each time the total is multiplied by 2.*

b. Clear the calculator and press 8 . Now press - 2 Press = three times. (Don't clear.)
Write your answers. 6, 4, 2
What is happening? *Each time 2 is subtracted from the total.*

d. Clear the calculator and press 3 2 ÷ 2 Now press = three times. (Don't clear.)
Write your answers. 16, 8, 4
What is happening? *Each time the total is divided by 2.*

Your calculator has an "automatic constant" feature. In each of the preceding examples "2" was the "constant" which was repeatedly added, subtracted, multiplied, or divided.

2. Count by 4's to 100 by pressing 4 + , then = as many times as necessary. Fill in the blanks.
4, 8, 12, 16, 20, 24, 28, 32, 36, 40, 44, 48, 52, 56, 60, 64, 68, 72, 76, 80, 84, 88, 92, 96, 100.

3. Display the first ten multiples of 6.
6, 12, 18, 24, 30, 36, 42, 48, 54, 60

• *Using the constant function*

Name _____

12

2-3

The Clever Calculator: Part III

More about constants and the calculator

1. Push ⑥ ⊞ ④ ⊟ . (Don't clear.)
Now push ④ ⊟ .
Answer: **10**
(Don't clear.) Push ⑦ ⊟ .
Answer: **13**
(Don't clear.) Push ⑧ ⊟ .
Answer: **14**
What's happening? *Each time 6 is added to the given number.*

2. Add 3 to each of the following:
(Push ③ ⊞ ⊟ first. Don't clear.)
a. 7 — **10**
b. 24 — **27**
c. 5 — **8**
d. 9 — **12**
e. 42 — **45**

3. Add 623 to each of:
a. 94 — **717**
b. 165 — **788**
c. 947 — **1570**
d. 382 — **1005**
e. 6493 — **7116**

4. Use a similar method to subtract 84 from each of the following:
(Push ① ⑤ ⑥ ⊟ ⑧ ④ ⊟ first, then ⑨ ③ ⑦ ⊟ ...)
a. 156 — **72**
b. 937 — **853**
c. 2518 — **2434**
d. 36,945 — **36,861**
e. 18,763 — **18,679**

5. Multiply 1.07 by each of:
a. 36 — **38.52**
b. 945 — **1011.15**
c. 167 — **178.69**
d. 25.84 — **27.6488**
e. 3765.93 — **4029.5451**

6. Divide each of the following by 63:
a. 441 — **7**
b. 227 — **3.603**
c. 63 — **1**
d. 97.4 — **1.546**
e. 2963.58 — **47.041**

• *Using the constant function*

Name _____

13

2-4

Who's a Cube?

Find out about squares and cubes

Squaring

Multiplying a number by itself (say, 6×6) is called "squaring" the number. Another way of writing 6×6 is 6^2 (read "six squared").

A fast method of squaring a number such as 6 on your calculator is by pressing the numeral key ⑥ , then the multiplication button ⊠ and then the equals sign ⊟ .
The answer 36 will appear.

Check all of your answers for the questions above, using the new method.

Find:
$6^2 = $ **36**, $9^2 = $ **81**, $8^2 = $ **64**, $41^2 = $ **1681**, $21^2 = $ **441**
$3^2 = $ **9**, $4^2 = $ **16**, $22^2 = $ **484**, $93^2 = $ **8649**, $88^2 = $ **7744**

Cubing

Multiplying a number by itself twice (say, $4 \times 4 \times 4$) is called "cubing" the number. Another way of writing $4 \times 4 \times 4$ is 4^3 (read "four cubed").

Find:
$4^3 = $ **64**, $9^3 = $ **729**, $2^3 = $ **8**, $17^3 = $ **4913**, $97^3 = $ **912,673**
$3^3 = $ **27**, $7^3 = $ **343**, $8^3 = $ **512**, $31^3 = $ **29791**, $100^3 = $ **1,000,000**

Using the information from "Squaring," can you find a fast method for cubing, using your calculator?

Fill in the table. Some answers are given as hints.

"Pushes"					
7	1	128	2187	16,384	78,125
6	1	64	729	4096	15,625
5	1	32	243	$4^5=1024$	3125
4	1	16	81	256	625
3	1	$2^3=8$	27	64	125
2	1	4	9	16	25
1	1	2	3	4	5
Start	1	1	1	1	1

• *Learning about exponents*
• *Optimizing the use of the calculator*

Name _____

14

2-6
What Multiplication Is: Part II
Which way is faster?

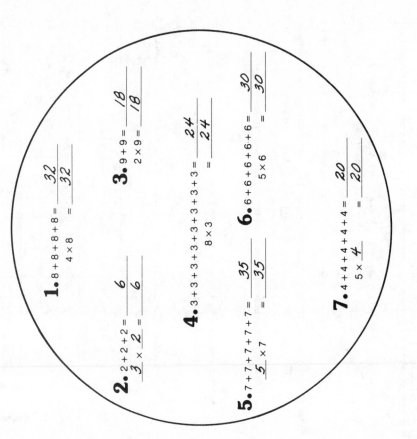

1. $8+8+8+8 =$ ___32___
   $4 \times 8 =$ ___32___

2. $2+2+2 =$ ___6___
   $3 \times 2 =$ ___6___

3. $9+9 =$ ___18___
   $2 \times 9 =$ ___18___

4. $3+3+3+3+3+3+3+3 =$ ___24___
   $8 \times 3 =$ ___24___

5. $7+7+7+7+7 =$ ___35___
   $5 \times 7 =$ ___35___

6. $6+6+6+6+6 =$ ___30___
   $5 \times 6 =$ ___30___

7. $4+4+4+4+4 =$ ___20___
   $5 \times 4 =$ ___20___

- Understanding multiplication
- Optimizing the use of the calculator

16

Name _____

2-5
What Multiplication Is: Part I
Look at the pictures two ways

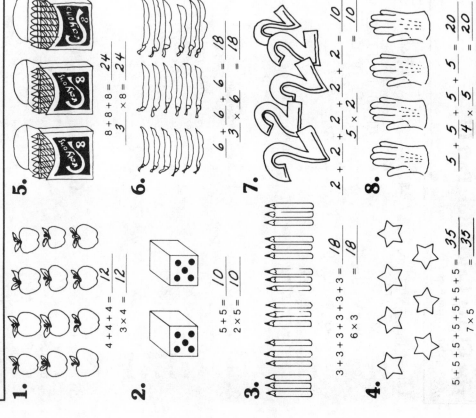

1. $4+4+4 =$ ___12___
   $3 \times 4 =$ ___12___

5. $8+8+8 =$ ___24___
   $3 \times 8 =$ ___24___

2. $5+5 =$ ___10___
   $2 \times 5 =$ ___10___

6. $6+6+6 =$ ___18___
   $3 \times 6 =$ ___18___

3. $3+3+3+3+3+3 =$ ___18___
   $6 \times 3 =$ ___18___

7. $2+2+2+2+2 =$ ___10___
   $5 \times 2 =$ ___10___

4. $5+5+5+5+5+5+5 =$ ___35___
   $7 \times 5 =$ ___35___

8. $5+5+5+5 =$ ___20___
   $4 \times 5 =$ ___20___

- Understanding multiplication
- Optimizing the use of the calculator

15

Name _____

2-8
The Important Order
What you put first makes a difference

a.

	4 + 5 × 3 = 19		6 + 7 × 4 = 34		2 + 5 × 4 = 22	
Agree	Disagree ✓	Agree	Disagree ✓	Agree	Disagree ✓

b.

5 × 3 + 4 = 9 /9	Correct answer: 19
7 × 4 + 6 = 34	Correct answer: 34
5 × 4 + 2 = 22	Correct answer: 22

c.

6 + 4 × 5 =	Rewritten: (if necessary) 4x5+6 Answer: 26	Correct answer: 26
9 + 3 × 5 =	Rewritten: (if necessary) 3x5+9 Answer: 24	Correct answer: 24
5 × 8 + 2 =	Rewritten: (if necessary) not necessary Answer: 42	Correct answer: 42

6 + 7 × 4 =	Rewritten: (if necessary) 7x4+6 Answer: 34	Correct answer: 34
8 × 3 + 7 =	Rewritten: (if necessary) not necessary Answer: 31	Correct answer: 31
8 × 7 + 4 =	Rewritten: (if necessary) not necessary Answer: 60	Correct answer: 60

d.

1. 2 + 7 × 5 = 37
2. 8 × 6 + 2 = 50
3. 9 × 5 + 4 = 49
4. 8 + 7 × 6 = 50

• *Discovering the need for operation-order rules*
• *Optimizing the use of the calculator*

18

2-7
The Commutative Property
How addition and multiplication are related

Complete. (Check your work with a calculator.)

1. 5 × 4 = 4 + 4 + 4 + 4 + 4 = 20
    4 × 5 = 5 + 5 + 5 + 5 = 20

2. 9 × 3 = 3 + 3 + 3 + 3 + 3 + 3 + 3 + 3 + 3 = 27
    27 = 6 + 6 + 6 = 6 × 3

3. 6 × 5 = 5 + 5 + 5 + 5 + 5 + 5 = 30
    5 × 6 = 6 + 6 + 6 + 6 + 6 = 30

4. 8 × 3 = 3 + 3 + 3 + 3 + 3 + 3 + 3 + 3 = 24
    3 × 8 = 8 + 8 + 8 = 24

5. 5 × 2 = 2 + 2 + 2 + 2 + 2 = 10
    2 × 5 = 5 + 5 = 10

6. 7 × 6 = 6 + 6 + 6 + 6 + 6 + 6 + 6 = 42
    6 × 7 = 7 + 7 + 7 + 7 + 7 + 7 = 42

7. 9 × 4 = 4 + 4 + 4 + 4 + 4 + 4 + 4 + 4 + 4 = 36
    4 × 9 = 9 + 9 + 9 + 9 = 36

8. 5 × 8 = 8 + 8 + 8 + 8 + 8 = 40
    8 × 5 = 5 + 5 + 5 + 5 + 5 + 5 + 5 + 5 = 40

Illustrate your answers with dot diagrams.

• *Reinforcing the concept of the commutative property*

17

Answer Key 63

2-10

Triple Choice

A real test of your estimating power

Select the smallest number, the middle number, and the largest number on each line.
Mark them A, B, and C respectively.
The first one is done for you.

	C	B	A
1.	C 698 + 544	B 328 + 443	A 107 + 343
2.	B 284 + 625	A 595 + 237	C 304 + 990
3.	C 487 + 964	A 107 + 748	B 432 + 622
4.	C 128 + 969	A 195 + 164	B 642 + 116
5.	C 340 + 865	A 354 + 307	B 283 + 505
6.	C 727 + 909	B 642 + 111	A 307 + 119
7.	A 238 + 109	B 116 + 369	C 430 + 277
8.	B 629 + 324	C 210 + 897	A 521 + 206
9.	C 435 + 802	A 382 + 208	B 705 + 284
10.	B 732 + 239	C 692 + 677	A 257 + 631
11.	C 564 + 916	A 125 + 533	B 616 + 212
12.	B 427 + 424	A 227 + 505	C 784 + 611

Now check with your calculator.

• Estimating three-digit numbers
• Rounding to the nearest ten or hundred

Name _____

2-9

Careful Comparisons

Just by looking, can you tell?

Insert <, =, or > to make the statements true.
Use your calculator to check.
The first problem is done for you.

Note

348 + 620	<	350 + 620
244 + 127	>	240 + 130
389 + 538	<	390 + 540
449 + 860	<	450 + 860
149 + 296	<	150 + 300
681 + 766	<	680 + 770
868 + 873	>	870 + 870
2801 + 1001	>	2800 + 1000
2945 + 9584	<	2950 + 9580
3595 + 5715	=	3600 + 5710
3087 + 9499	<	3090 + 9500
4934 + 1536	=	4930 + 1540

• Estimating in addition
• Seeing number relations

Name _____

2-11
The Educated Guess
How close can you come how fast?

Record your guess for the sum of the numbers in each circle.
Then check with your calculator.

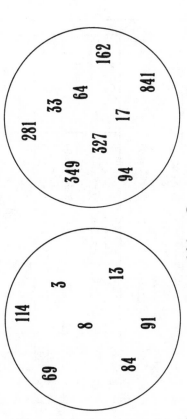

69   114   3   13
   8
84   91

357   63   39   94
   3   242
   10   5   16

1. Guess: _____ Exact sum: *382*
Was your guess too high or too low?

3. Guess: _____ Exact sum: *829*
Was your guess too high or too low?

281   33   64   162
349   327   17
   94   841

23   105   18
   129   48
84   120   273

2. Guess: _____ Exact sum: *2168*
Was your estimate too high or too low?

4. Estimate: _____ Exact sum: *800*
Was your estimate too high or too low?

• *Estimating in addition*

Name _____

21

2-12
Tri-Option
Biggest number gets a circle

Use rounding to select the largest number on each line.
Circle your answer. Do not calculate.

53 × 7	50 × 8	(19 × 37)
38 × 29	23 × 42	(91 × 13)
47 × 34	90 × 16	(25 × 70)
89 × 41	60 × 60	(75 × 50)
25 × 49	(75 × 20)	89 × 15
23 × 94	45 × 45	(53 × 42)
47 × 61	21 × 94	(53 × 57)
54 × 54	(49 × 74)	68 × 29
84 × 43	62 × 63	(45 × 90)
56 × 72	(65 × 65)	45 × 85
(61 × 70)	29 × 99	43 × 97
44 × 39	89 × 17	(42 × 48)

Check your answers with your calculator.

• *Estimating in multiplication*
• *Rounding to the nearest ten*

Name _____

22

2-14
Home, Home on the Range: Part II

Check your estimating habits

a. Write your estimate for the high and low values between which the answer will fall. Make the difference between the high and low as small as you dare.

b. Check by finding the exact answer with your calculator.

c. Write down the difference between the low value and the exact answer. Do the same for the high value and the exact answer.

d. Add columns (d) and (e) to find your total score.

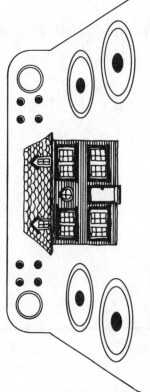

	(a) >	(b) <	(c) Exact answer	(d) |c − a|	(e) |b − c|
23 × 19	400	460	437	37	23
51 × 37			1887		
49 × 36			1764		
54 × 48			2592		
96 × 42			4032		
51 × 36			1836		
94 × 94			8836		
83 × 68			5644		
45 × 45			2025		
73 × 81			5913		
61 × 93			5673		

Score:

Estimates may vary.

Name

• *Estimating in multiplication*
• *Using the concept of a range*
• *Judging error in the estimation process*

24

2-13
Home, Home on the Range: Part I

You choose the challenge

a. Write your estimate for the range into which the answer will fall. Make the range as small as you dare.

b. Check by finding the exact answer with your calculator.

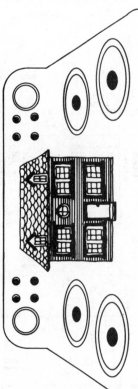

	Step (a) >	Step (b) Exact answer	(a) <
6 × 429	2 400	2 574	3 000
4 × 826		3304	
9 × 454		4086	
7 × 621		4347	
9 × 844		7596	
3 × 251		753	
473 × 8		3784	
261 × 4		1044	
356 × 9		3204	
257 × 3		771	
641 × 6		3846	
254 × 8		2032	

Estimates may vary. Encourage students to narrow the range.

Name

• *Estimating in multiplication*
• *Developing the concept of a range*

23

2-15

Home, Home on the Range: Part III

How well do you estimate bigger numbers?

a. Write your estimate for the range into which the answer will fall.
   Make the range as small as you dare.
b. Check by finding the exact answer with your calculator.
c. Write down the difference between the low of the range and the exact answer.
   Do the same for the high of the range and the exact answer.
d. Add columns (d) and (e) to find your score.

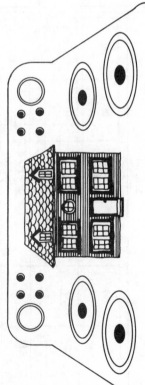

	(a) >	(b) <	(c) Exact answer	(d) |c − a|	(e) |b − c|
60 × 437	25,500	27,000	26,220	720	780
40 × 965			38,600		
70 × 259			18,130		
80 × 496			39,680		
70 × 654			45,780		
137 × 90			12,330		
446 × 20			8920		
50 × 475			23,750		
625 × 40			25,000		
343 × 20			6860		
545 × 60			32,700		

Score:

Estimates may vary.

Name _____

• Estimating in multiplication
• Using the concept of a range
• Judging error in the estimation process

2-16

Home, Home on the Range: Part IV

Try to get these even closer!

a. Write your estimate for the range into which the answer will fall.
   Make the range as narrow as you can.
b. Check by finding the exact number with your calculator.
c. Write the difference between the low of the range and the exact answer.
   Do the same for the high of the range and the exact answer.
d. Add columns (d) and (e) to find your score.

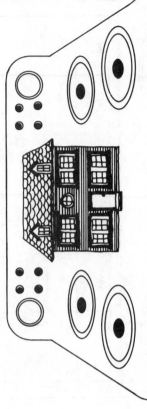

	(a) >	(b) <	(c) Exact answer	(d) |c − a|	(e) |b − c|
300 × 64	18,000	19,500	19,200	1200	300
700 × 85			59,500		
900 × 61			54,900		
600 × 43			25,800		
700 × 29			20,300		
49 × 500			24,500		
62 × 700			43,400		
91 × 600			54,600		
37 × 400			14,800		
73 × 800			58,400		
34 × 900			30,600		

Score:

Estimates may vary.

Name _____

• Estimating in multiplication
• Using the concept of a range
• Judging error in the estimation process

2-18
Home, Home on the Range: Part VI
Try for an even lower score!

a. Write your estimate for the low and high values between which the answer will fall. Make the difference between these values as small as you can.
b. Check by finding the exact answer with your calculator.
c. Write the difference between your low estimate and the exact value. Do the same for your high estimate.
d. Add columns (d) and (e) to find your score.

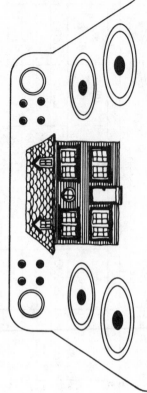

	(a) >	(b) <	(c) Exact answer	(d) |c − a|	(e) |b − c|
7 × 61			427		
30 × 84			2520		
97 × 40			3880		
385 × 174			66,990		
80 × 672			53,760		
37 × 42			1554		
800 × 258			206,400		
474 × 800			379,200		
60 × 493			29,580		
300 × 296			88,800		
897 × 6			5382		

Score: _____ *Estimates may vary.*

• *Estimating large products*
• *Using the concept of a range*
• *Judging error in the estimation process*

Name _____

28

2-17
Home, Home on the Range: Part V
The numbers are larger but don't give up

a. Write your estimate for the range into which the answer will fall. Make the range as narrow as you dare.
b. Check by finding the exact answer with your calculator.
c. Write down the difference between the low of the range and the exact answer. Do the same for the high of the range and the exact answer.
d. Add columns (d) and (e) to find your score.

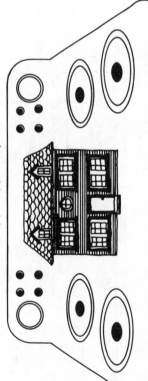

	(a) >	(b) <	(c) Exact answer	(d) |c − a|	(e) |b − c|
174 × 61	10,000	11,000	10,614	614	386
37 × 451			16,687		
26 × 934			24,284		
97 × 614			59,558		
298 × 47			14,006		
361 × 94			33,934		
369 × 247			91,143		
451 × 361			162,811		
258 × 414			106,812		
309 × 646			199,614		
747 × 818			611,046		

Score: _____ *Estimates may vary.*

• *Estimating large products*
• *Using the concept of a range*
• *Judging error in the estimation process*

Name _____

27

2-19

Somewhere In-between

Several answers can be equally right!

Fill in the blanks to get an answer in the given range.
The first one is done for you.

Multiplication		Range
34 ×	2.5	70 → 90
56 ×	6	300 → 350
23 ×	27	600 → 650
17 ×	30	500 → 550
98 ×	5	450 → 500
12 ×	9	100 → 120
22 ×	23	480 → 520
38 ×	7	220 → 280
64 ×	15	950 → 1000
31 ×	20	600 → 650
75 ×	9	650 → 700
52 ×	16	820 → 880
49 ×	17.5	840 → 880
61 ×	3.5	200 → 240
26 ×	6	140 → 170
33 ×	26	830 → 870
41 ×	14	570 → 610
24 ×	37	860 → 900
57 ×	4.5	250 → 280
19 ×	26	475 → 500

Estimates may vary.

· Estimating in multiplication
· Estimating in division

2-20

A Santa Claus Riddle

What advice does Santa Claus give to farmers?

Answer

Color in the division expression with the biggest answer for each row (a, b, c, etc.).
In rows c, e, and g, color in the 2 largest expressions for the row.
Estimate, don't work them out!
You may check with your calculator.
Row "a" is done for you.

Rotate the page to find the answer!

a	34 ÷ 5		69 ÷ 7		167 ÷ 9		
b	45 ÷ 7		129 ÷ 12		12 ÷ 3		
c	555 ÷ 25		26 ÷ 8		29 ÷ 4		
d	67 ÷ 6		337 ÷ 19		37 ÷ 8		
e	57 ÷ 1		34 ÷ 3		918 ÷ 11		
f	118 ÷ 8		58 ÷ 2		465 ÷ 64		
g	834 ÷ 92		52 ÷ 9		237 ÷ 6		
h	50 ÷ 7		91 ÷ 2		831 ÷ 36		
i	92 ÷ 3		228 ÷ 17		16 ÷ 3		

HO,HO,HO (HOE,HOE,HOE)

· Estimating in division

2-25

Roundabout Route to a Square Root

A memory-and-estimation-challenge   (for 2 players)

Equipment

Calculators
Set of cards (below)

Instructions

1. Cut out the cards, shuffle them, and place them face down in a pile.
2. Player A turns over the top card. He or she estimates the square root of the number on the card, and checks with the calculator (b × b = ?).
3. Player B may watch A's attempt. If A's guess is wrong, B takes a guess. Players continue, taking turns "narrowing in" on the target.
4. The player who finds the correct square root keeps the card.
5. Play continues until all cards have been used. The player with the most cards wins.

52	*556*	*857*	*78*
2704	309,136	734,449	6084
89	*337*	*412*	*44*
7921	113,569	169,744	1936
81	*68*	*788*	*584*
6561	4624	620,944	341,056
36	*43*	*777*	*789*
1296	1849	603,729	622,521
258	*56*	*97*	*124*
66,564	3136	9409	15,376

Name _____

35

• *Estimating in multiplication or division*
• *Using square roots*

Sheet 2-21 through Sheet 2-24
No specific answers.

2-29
A Fraction by Another Name

What can you call these spaces?

Assign a common fraction to the shaded part in each drawing.
Then use your calculator to find the corresponding decimal fraction.

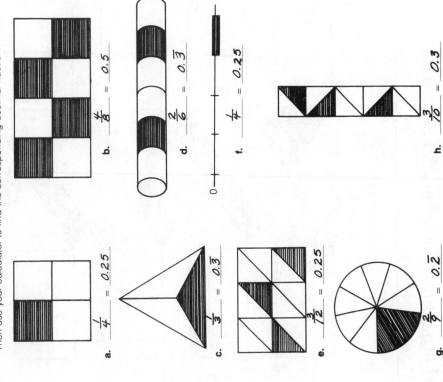

a. $\frac{1}{4}$ = 0.25

b. $\frac{4}{8}$ = 0.5

c. $\frac{1}{3}$ = $0.\overline{3}$

d. $\frac{2}{6}$ = $0.\overline{3}$

e. $\frac{3}{12}$ = 0.25

f. $\frac{1}{4}$ = 0.25

g. $\frac{2}{9}$ = $0.\overline{2}$

h. $\frac{3}{10}$ = 0.3

Name _____

• Understanding the concept of fractions
• Converting fractions to decimal form

Sheet 2-26 through Sheet 2-28
No specific answers.

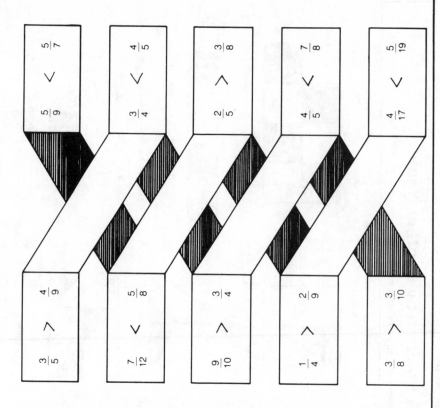

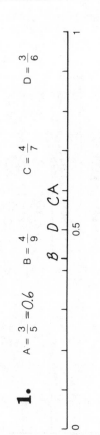

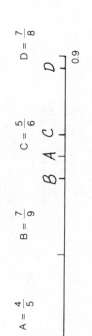

2-31
Larger Fractions, Smaller Fractions
A matter of comparing

Put the correct sign (>, <, or =) between each pair of fractions.
Check by converting to decimals, using your calculator.

$\frac{3}{5} > \frac{4}{9}$ $\frac{5}{9} < \frac{5}{7}$

$\frac{7}{12} < \frac{5}{8}$ $\frac{3}{4} < \frac{4}{5}$

$\frac{9}{10} > \frac{3}{4}$ $\frac{2}{5} > \frac{3}{8}$

$\frac{1}{4} > \frac{2}{9}$ $\frac{4}{5} < \frac{7}{8}$

$\frac{3}{8} > \frac{3}{10}$ $\frac{4}{17} < \frac{5}{19}$

• Comparing fractions
• Converting fractions to decimal form

Name _____

2-30
The Fraction Line
How do you tell where they fall?

Place these points on the number line.
The first one is done for you.

1.

$A = \frac{3}{5} = 0.6$ $B = \frac{4}{9}$ $C = \frac{4}{7}$ $D = \frac{3}{6}$

B   D   CA

0 ————————— 0.5 ————————— 1

2.

$A = \frac{4}{5}$ $B = \frac{7}{9}$ $C = \frac{5}{6}$ $D = \frac{7}{8}$

B   A   C   D

0.6 ————————————— 0.9

3.

$A = \frac{7}{12}$ $B = \frac{3}{8}$ $C = \frac{1}{3}$ $D = \frac{1}{4}$

D   C   B   A

0 ————————— 0.5 —————————

• Converting fractions to decimal form
• Comparing fractions

Name _____

2-32
A Fraction by Another Name: II

Each fraction has a decimal "alias"

Find the decimal name for each number.
Carry out the division to as many decimal places as your calculator allows.

	Decimal		Decimal		Decimal		Decimal		Decimal
$\frac{1}{1}$	1	$\frac{1}{5}$	0.2	$\frac{1}{9}$	0.$\overline{1}$	$\frac{1}{13}$	0.076923	$\frac{1}{17}$	0.0588235
$\frac{1}{2}$	0.5	$\frac{1}{6}$	0.1$\overline{6}$	$\frac{1}{10}$	0.1	$\frac{1}{14}$	0.0714285	$\frac{1}{18}$	0.0$\overline{5}$
$\frac{1}{3}$	0.$\overline{3}$	$\frac{1}{7}$	0.142857I	$\frac{1}{11}$	0.$\overline{09}$	$\frac{1}{15}$	0.06$\overline{6}$	$\frac{1}{19}$	0.0526315
$\frac{1}{4}$	0.25	$\frac{1}{8}$	0.125	$\frac{1}{12}$	0.083$\overline{3}$	$\frac{1}{16}$	0.0625	$\frac{1}{20}$	0.05

Use the table to complete the following.

From the Table		By Multiplication		By Division		
$\frac{1}{4}$	0.25	$\frac{3}{4} = (3\times0.25)$	0.75	$3\div4$	0.75	
$\frac{1}{6}$	0.1$\overline{6}$	$\frac{2}{6} = (2\times0.1\overline{6})$	0.3333332	$2\div6$	0.$\overline{3}$	*
$\frac{1}{8}$	0.125	$\frac{7}{8} = (7\times0.125)$	0.875	$7\div8$	0.875	
$\frac{1}{3}$	0.$\overline{3}$	$\frac{2}{3} = (2\times0.\overline{3})$	0.$\overline{6}$	$2\div3$	0.$\overline{6}$	
$\frac{1}{9}$	0.$\overline{1}$	$\frac{5}{9} = (5\times0.\overline{1})$	0.$\overline{5}$	$5\div9$	0.$\overline{5}$	
$\frac{1}{10}$	0.1	$\frac{4}{10} = (4\times0.1)$	0.4	$4\div10$	0.4	
$\frac{1}{12}$	0.083$\overline{3}$	$\frac{7}{12} = (7\times0.083\overline{3})$	0.583331	$7\div12$	0.5833331	*
$\frac{1}{18}$	0.0$\overline{5}$	$\frac{11}{18} = (11\times0.05\overline{5})$	0.611105	$11\div18$	0.611111	*

What is unusual about the answers to the questions marked with an asterisk (*)? Explain. *See Teacher's Guide.*

- *Converting fractions to decimal form*
- *Building a table*
- *Comparing processes for determining decimal equivalents*

Name _____

2-33
Guess How It Goes From Here

Look closely at what repeats

1. Use your calculator to find the decimal equivalent for the first three fractions in the group. Then guess the others.

Calculate:

$\frac{1}{3}$ = 0.3333333

$\frac{1}{33}$ = 0.0303030

$\frac{1}{333}$ = 0.0030030

Guess:

$\frac{1}{3333}$ = 0.00030003

$\frac{1}{33,333}$ = 0.000030003

Check with your calculator.

2. Calculate:

1 × 1 = 1

11 × 11 = 121

111 × 111 = 12,321

Guess:

1111 × 1111 = 1,234,321

Check with your calculator.

These products go beyond the calculator display. Predict the answers. How can they be checked?

11,111 × 11,111 = 123,454,321

111,111 × 111,111 = 12,345,654,321

3. Calculate:

9 × 9 = 81

99 × 99 = 9801

999 × 999 = 998,001

Guess:

9999 × 9999 = 99,980,001

Check with your calculator.

These products go beyond the calculator display. Can you predict the answers? How can they be checked?

99,999 × 99,999 = 9,999,800,001

999,999 × 999,999 = 999,998,000,001

- *Detecting and completing patterns*
- *Using intuition*

Name _____

2-34

Pattern Discovery

Discovering your own with a number array

Consider the array below.

1	2	3	4	5	6	7	8	9	10
11	12	13	14	15	16	17	18	19	20
21	22	23	24	25	26	27	28	29	30
31	32	33	34	35	36	37	38	39	40
41	42	43	44	45	46	47	48	49	50
51	52	53	54	55	56	57	58	59	60
61	62	63	64	65	66	67	68	69	70
71	72	73	74	75	76	77	78	79	80
81	82	83	84	85	86	87	88	89	90
91	92	93	94	95	96	97	98	99	100

1. Compare the pair 34, 45 with the pair 44, 35.

a. Add: 34 + 45 = _79_

44 + 35 = _79_

What do you notice about the answers?

They're the same.

Is that finding true for any two such pairs? _Yes_

b. Subtract: 45 − 34 = _11_

44 − 35 = _9_

How do the answers differ? *by 2*

Is that true for any two such pairs? _Yes_

c. Multiply: 34 × 45 = _1530_

35 × 44 = _1540_

How do the answers differ? *by 10*

Is that true for any two such pairs? _Yes_

2. There are many other patterns you can find.

Describe some of the patterns you found.

Do they hold true if the array is extended to include negative numbers?

See hints drawn on array.

3. What happens to your pattern if the array is changed as follows?

1	2	3	4	5	6
7	8	9	10	11	12
13	14	15	16	17	18
19	20	21	22	23	24

+, the same

−, always 7,5 (differ by 2)

×, differ by 6

• *Detecting and completing patterns*
• *Drawing conclusions*

Name _____

2-36
Stocking Up
Estimate the costs of this campout

The girl scouts are going camping. Three girls have been assigned to buy the supplies.
Estimate how much money each scout should be given for supplies.
Then use these estimates to estimate the total amount of money needed.
Finally, check all figures with your calculator.

Peggy's list

		Estimate	Actual
30 oranges	@ $0.15 each	$ 4.00	$4.50
6 heads of lettuce	@ 0.39 each		2.34
4 bags of apples	@ 1.29 each		5.16
5 bunches of carrots	@ 0.49 each	*Estimates*	2.45
3 watermelons	@ 1.98 each	*will vary.*	5.94
2 baskets of tomatoes	@ 1.49 each		2.98
	Total		$23.37

Kathleen's list

		Estimate	Actual
12 cans of beans	@ $0.32 each	$ 3.50	$ 3.84
4 cans of tomato soup	@ 0.26 each		1.04
6 cans of mushroom soup	@ 0.29 each		1.74
4 jars of peanut butter	@ 1.24 each	*Estimates*	4.96
2 jars of strawberry jam	@ 1.29 each	*will vary.*	2.58
4 packages of hot dogs	@ 1.12 each		4.48
	Total		$18.64

Patti's list

		Estimate	Actual
2 boxes of milk powder	@ $2.25 each		$4.50
2 boxes of soap	@ 1.49 each		2.98
6 loaves of bread	@ 0.47 each		2.82
3 boxes of breakfast cereal	@ 1.12 each	*Estimates*	3.36
4 packages of buns	@ 0.59 each	*will vary.*	2.36
2 bags of cookies	@ 1.38 each		2.76
	Total		$18.78

Estimated total sum of money needed for the 3 girls: *will vary*
Actual total sum of money needed for the 3 girls: $60.79

· *Estimating in multiplication*
· *Building consumer math skills*

46

2-37
An Expensive House!
See how the materials add up

Mary and her parents are building a playhouse.
As you can see, they have already purchased many of the materials.
Estimate the total cost of the materials so far.

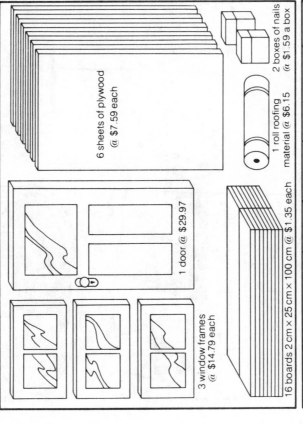

6 sheets of plywood @ $7.59 each

1 door @ $29.97

3 window frames @ $14.79 each

16 boards 2 cm × 25 cm × 100 cm @ $1.35 each

1 roll roofing material @ $6.15

2 boxes of nails @ $1.59 a box

Item	Estimate	Actual	Item	Estimate	Actual
Roofing		$6.15	Plywood		$45.54
Windows		$44.37	Nails		$3.18
Boards		$21.60	Door		$29.97

Estimates will vary.

Estimated Total _____
Actual Total $150.81

Use your calculator to check your answers.

· *Estimating in multiplication*
· *Gathering information*
· *Building consumer math skills*

47

2-39
The World's Largest Swimming Pool
Figure areas and volumes

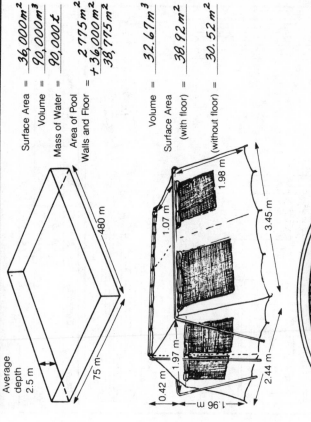

Average depth 2.5 m
480 m
75 m

Surface Area = *36,000 m²*
Volume = *90,000 m³*
Mass of Water = *90,000 t*
Area of Pool
Walls and Floor = *2775 m²*
+ 36,000 m²
38,775 m²

1.98 m
1.07 m
3.45 m
1.97 m
2.44 m
0.42 m
1.96 m

Volume = *32.67 m³*
Surface Area
(with floor) = *38.92 m²*
(without floor) = *30.52 m²*

Royal Regency
round model

3.16 m
0.97 m

Capacity = _30,430_ L

- *Working with area and volume*
- *Gathering information from drawings*
- *Working with pi*

Name _____

49

2-38
Finding the Best Buy
Think first before you spend

Buying a Freezer

$339.98
1.87 m
0.78 m
0.82 m

$349.98
1.08 m
0.82 m
1.93 m

Which freezer is the best buy? Why? *You get almost 5dm³/$1, compared with 3.5 dm³/$1 for the upright.*

Buying a Mower

$79.95

$89.95

Width of cut: 50 cm
Width of cut: 45 cm

The $79.95 mower will cut more grass per dollar (the cost per cm of blade is less: $1.79/cm vs $1.80/cm).

Which mower is the best buy? *The $79.95 one.*
Why?

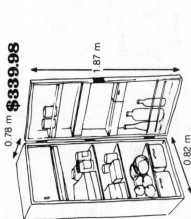

- *Working with volume and ratio*
- *Calculating cost per unit*

Name _____

48

2-43

Superstar Publicity

Figure out the price of fame

Singer I. M. Great publicly promised each of her fans an autographed photo of herself. The response was bigger than expected: all 220,000,000 people in the U.S. wrote for a picture. (Even her own brother!)

Estimate, then calculate, the number of hours I. M. will need to sign all photographs, if she signs 1 picture every 3 s (3 seconds).

Estimated number of hours: *Answers will vary.*

Actual number of hours: *183,333.33h*

Estimate, then calculate, the number of 40-h work weeks I. M. will have to spend signing the pictures.

Estimated number of weeks: *Answers will vary.*

Calculated number of weeks: *4583.3332 weeks*

Be careful not to go beyond the capacity of your calculator.

I. M. starts signing at age 18. Each year she takes 2 weeks off for vacation. How old will she be when she finishes autographing the photos?

Age: *109 years*   *Danger!*

There are 20 photographs in a thickness of 1 cm. How high is I. M.'s pile of pictures, in kilometers? *110 km*

Each photograph has a mass of 5 g. Estimate, then calculate, the number of large 5 t trucks required to take the photographs to the post office.

Estimated number of 5 t trucks: *Answers will vary.*

Actual number of 5 t trucks: *220*

Be careful!

Name _____

• *Working with large numbers*

53

Sheet 2-40
Answers will vary.

Sheet 2-41
Answers will vary.

Sheet 2-42
In Problem 11, the volume of the pool is 93,974 liters. All other answers will vary. For Problem 12, students can consider the effect of evaporation on a pool of this size.

2-44
The Point in the Right Place
Size up about what's right

1. Multiply:

46	37	924	65	248
24	68	37	89	37
1104	**2156**	**34,188**	**5785**	**9176**
659	514	258	383	296
236	32	467	25	37
155,524	**16,448**	**120,486**	**9575**	**10,952**
1465	1543	284	495	2467
29	614	16	109	9
42,485	**947,402**	**4544**	**53,955**	**22,203**
454	679	6794	4645	3882
43	28	3144	3175	916
19,522	**19,012**	**21,360,336**	**14,747,875**	**3,555,912**

2. Multiply:

4.6	3.7	9.24	0.65	0.248
2.4	68	3.7	0.89	0.37
11.04	**215.6**	**34.188**	**0.5785**	**0.09176**
6.59	51.4	0.258	383	2.96
2.36	3.2	467	2.5	37
15.5524	**164.48**	**120.486**	**957.5**	**109.52**
146.5	1543	0.284	0.495	2467
0.29	0.614	1.6	1.09	0.09
42.485	**947.402**	**0.4544**	**0.53955**	**222.03**
454	67.9	67.94	4645	0.3882
4.3	28	3144	3.175	916
1952.2	**1901.2**	**213,603.36**	**14,747.875**	**355.5912**

• *Working with rules for decimal placement*
• *Detecting and explaining patterns*

Name _____

54

2-45
Solving Sentences: Part I
An educated guess may do it!

Guess which number goes into the box to make the sentence true. Check with your calculator. If the guess is not quite right, cross out the answer and try again, and again, until it fits exactly. An example is done for you.

4 × [~~8~~ ~~7~~ 4.5✓] + 5 = 23

← Put your first guess here.

← Write your second guess here.

Stop guessing when the answer fits.

Tries will vary.

1. 2 × [4.5] + 7 = 16
2. 5 × [3.4] − 4 = 13
3. 10 × [3.3] + 4 = 37
4. 5 × [8.4] − 13 = 29
5. 6 × [11] + 14 = 80
6. 4 × [6.5] − 12 = 14
7. 10 × [5.6] − 6 = 50
8. 5 × [7.6] − 3 = 35
9. 3 × [12] + 5 = 41
10. 4 × [8.75] − 7 = 28
11. 10 × [4.5] + 3 = 48
12. 4 × [7.25] − 4 = 25

• *Solving equations with one variable*
• *Applying estimating skills*

Name _____

55

2-46
Solving Sentences: Part II

Get the answer to this riddle:

Why is a teacher like a candy maker?

clues

1. Find a number (you'll need decimals) that makes each number sentence correct.
2. Write the letter of the problem above the correct answer at the bottom of the page. The first one is done for you.

K: $5 \times \boxed{3.4} + 3 = 20$

A: $(5 + \boxed{2.2}) \times 5 = 36$

U: $\dfrac{3 \times \boxed{2.6}}{2} = 3.9$

S: $4 \times \boxed{8.5} + 5 = 39$

L: $\boxed{3.1}^2 - 0.61 = 9$

O: $\dfrac{5 \times \boxed{3.7} + 1.5}{4} = 5$

T: $6 \times \boxed{4.5} - 9 = 18$

E: $\boxed{2.5}^2 + 0.75 = 7$

H: $4 \times \boxed{3.5} + 3 = 17$

C: $(2.05 + \boxed{0.2}) \times 8 = 18$

L: $5 \times \boxed{3.2} - 3 = 13$

A: $(4.3 + \boxed{0.3}) \times 5 = 23$

answer
They
both

U	S	E	C	H	A	L	K	A	L	O	T	!
2.6	8.5	2.5	0.2	3.5	2.2	3.2	3.4	0.3	3.1	3.7	4.5	

Name _____

• Solving equations with one variable
• Applying estimating skills

56

Sheet 2-47
Answers will vary.

Teacher's Notes

Teacher's Notes

Teacher's Notes

Teacher's Notes

Teacher's Notes

Teacher's Notes

Teacher's Notes